America's Two-Headed Pig

By Leah Dunham

With contributions by Arthur Dunham, DVM

www.americastwoheadedpig.com

This book is a Wholesome Note Publication.

Cover Art by Carol Ewing.

Cover Design by Lawson Wulsin.

ISBN: 0989484009
ISBN-13: 978-0989484008

DEDICATION

This book is dedicated to Benjamin Ewing.

CONTENTS

ACKNOWLEDGMENTS

I would like to acknowledge everyone in Delaware County, Iowa who has ever attempted to understand my father's hour-long diatribes about crop chemicals and animal diseases. I hope that this book helps clarify a few things!

AN INTRODUCTION

My dad is a large animal veterinarian. When I was in first grade, he brought home a two-headed pig. After acknowledging that it would never survive, his client was more than willing to give it up. My siblings and I cared for this curious creature until its premature death.

I convinced myself that with proper care the piglet might survive. Everything about it seemed so fascinating and miraculous; how could it not? But alas, each time we attempted to feed one head, the other head ran interference. Just as one pink mouth met the bottle, the other decided it needed the very same thing. Our little pig's heads bobbed back and forth. The poor creature never achieved its ultimate goal: to nourish itself.

As it turns out, the only way to care for a two-headed pig is to bottle feed both heads simultaneously. Still, this was not enough—it was simply too deformed to survive. Unable to make up its minds, our pig was never able to obtain enough nutrients. It died within a week. We had all grown rather attached and were saddened by its sudden departure.

When I reflect on modern American agriculture, I occasionally think back to our two-headed pig. The heads influencing American farms seem similarly confused, often pressing policies that help large monoculture operations while hurting small, diverse farms, one head attempting to receive nourishment at the cost of another. These policies have deformed our agricultural system to such an extent that they jeopardize the quality of the natural resources we rely on for food. As a result, we undermine one of our farmers' ultimate goals: to feed hungry citizens nourishing food.

With such policies in place, industrial ag conglomerates and chemical companies will continue to devour our global nutritive supply, replacing land resources, heirloom seeds, and jobs with a handful of ag operations and bioengineered monocultures. These streamlined operations will be completely dependent on manufactured fertilizers, seeds, and pesticides. While these operations will produce an abundance of food, safe, nourishing food, produced by prosperous communities, will become harder to find. In fact, as many food activists have reported, global citizens are already experiencing food safety issues due to misinformed ag policies.

The most obvious food safety issues associated with industrial ag are the outbreaks of food-borne illness that we all hear about in the news. But there are signs that this way of growing crops is damaging our health in other, far more insidious ways—among them, by damaging the health of animals raised for food.

In this book I demonstrate how seemingly disparate problems in the fields of vet medicine, human medicine, and plant medicine stem from common roots within the American food supply. Each chapter is organized around a particular diagnosis. Within each chapter, I explore how an animal illness has been complicated by two-headed science. As I am about to explain in Diagnosis One, two-headed science is science that has been deformed and corrupted by vested interests. I also use stories from my personal experiences on Iowa livestock farms to illustrate the numerous connections between these illnesses and America's misinformed farm policies.

My father has pored over thousands of research papers in attempts to remedy the underlying causes of the illnesses described in this book.

His work has embodied a commitment to healthy lands, creatures, and farms, as well as the hard work necessary to sustain them. He is listed as a contributor to the book because his experiences and research have informed much of what I've written. After years of listening to him talk about his attempts to solve reoccurring health problems, I realized that most people don't have a clue as to how modern disease complexities affect farm animals. We both hope that this book will help all medical professionals, farmers, and consumers better address the true roots of various medical conditions, including nutrient deficiencies, clostridial infections, diabetes, and Parkinsons disease.

That said, my father does not agree with my entire perspective. Our differences lie largely in one area. While I am an idealist who thinks that pigs and cows should spend some of their days on pasture, my father thinks that a shift away from Confined Animal Feeding Operations (CAFOs) toward pastured operations will take more time and effort than the average American is willing to put forth. In other words, he is convinced that confinement units, in which thousands of animals are raised indoors for their entire lives, are here to stay. Though he agrees that CAFOs do increase the prevalence of certain diseases, he is primarily interested in remedying the ways modern feed products influence animal health. This is because he thinks that the animal feed supply will be easier to modify than the entire CAFO model. Ultimately, he would like to see all of his clients, both confinement operators and farmers, achieve better animal health measures. To do so, he believes that crop farmers will have to reduce their reliance on crop chemicals, antibiotics, and gene technologies, because these feed inputs interfere with optimal health.

While I fully support these goals, I believe that both plant and animal farmers will need to make changes to better address the health concerns described in this book. Because we disagree on this matter, we have had several debates to sort out what should be included in the following pages. While reading Diagnosis One, for example, my father worried that if I present too many criticisms of modern CAFO systems, researchers and vets might not read the following chapters because they might take these criticisms personally. While I understand where he's coming from, I wanted to paint an accurate portrait of all of the factors that have historically influenced disease prevalence. Leaving CAFOs out of the conversation didn't seem like an honest option. In the end, we decided that we should expose as many faults with current plant and animal models as possible and let readers sort through their own biases. If you can't stand a chapter, write us a letter explaining why. We'd love to learn something new about a disease we're attempting to address. This is why we decided to share the following information in the first place.

Fortunately, despite this difference in perspective, we found that we could agree on the nature of the illnesses themselves. I did my best to represent my father's observations in an accurate and meaningful way while staying true to my own values and goals. Armed with increased awareness of how plant, animal, and human ailments interrelate, consumers, policymakers, and medical professionals can make healthier decisions for our nation's families. This book serves as further evidence that it is time to move away from unsustainable ag practices toward more diverse, ecology-based farming methods.

DIAGNOSIS 1: VITAMIN D DEFICIENCY

When my father and his partner opened their veterinary business in 1974, Delaware County, IA was the leading, most dense hog-producing county in the nation. Out of 1500 farms, 1158 sold pigs. When my father wasn't attending pigs, he was often assisting with matters on more than 800 cattle farms, either beef or dairy, scattered throughout the county.[1] In addition to cattle and hogs, he treated chickens, sheep, goats, dogs, cats, llamas, emus, and the occasional two-headed pig.

According to the last ag census, the number of Delaware County farms selling pigs had dropped to 285. This number is misleading in itself as many of these pigs are owned by the same vertically integrated agribusiness. In the eighties and nineties Delaware County farms consolidated to remain competitive with new, improved factory farms churning out ever-growing volumes of cheap dairy, beef, and pork products. Those who couldn't expand gave up farming altogether or signed contracts with larger agribusinesses to relieve growing economic burdens.

Today, my father groans every time he reads news highlighting improved efficiency and productivity measures in the animal ag industry. While pig sales in Delaware County are about 35% higher than they were in 1974, the number of hog farms has actually decreased by 75%. The bulk of hog sales are conducted by large confinement operations.[1,2] Most of these family-run operations are bound to decision-making processes that are no longer their own.

Approved mergers and poorly informed policies have made crop farming and confinement units the safest options for remaining Delaware County farmers. The way to make money here is clear: you can plant corn, raise soybeans, or pack thousands of animals into one confinement facility. Do anything that is not backed by a contract or supported by federal policies and you might as well commit yourself to a life of poverty.

Rural landscapes haven't always been so discriminating. As a kid, my father eagerly pointed out the beef and dairy cows, pigs, horses, alfalfa, oats, rye, and wheat on various drives throughout Northeastern Iowa. On a recent two-hour drive through this same region, we drove by miles and miles of corn and soybeans occasionally and excitedly acknowledging a farm that still had pastured animals. More often than not, we noticed another acre planted to the same, farmers trying to eke out return from every inch of available land. Few farmers work against the grain sparing acreage for something else provided the profit it will cost them.

We search for evidence that consumers' concerns for rural environments and livelihoods are being realized—aspects of agriculture beyond this tightly controlled world. While a handful of farmers stubbornly hold on, we drive for miles without finding much evidence. My father simultaneously faces the hard reality that his veterinary career is not as secure as he once thought.

His concerns are justifiable. A practice that used to support three veterinarians now barely supports two. As my father and his partner contemplate retirement, the future of their business remains uncertain. When they originally invested in their practice, they imagined passing their business and equipment on to three new veterinarians. Today my father wonders if they'll eventually need to merge with another practice in another county to remain profitable.

My father is not alone. Like many livestock producers, large animal veterinarians are struggling to make ends meet throughout much of rural America. According to a recent National Research Council report, the decline in demand for rural veterinary services coincides with changes in ag industry structure. One of the chief causes mentioned in the report is the increasing number of animals on a decreasing number of farm-operations. The NRC also points out that large producers tend to focus more on routine herd health, like vaccinations, dehorning, and shots, rather than regular care. Farm operators delegate such procedures to farm hands rather than trained and licensed professionals.

The shrinking demand for professional large animal veterinarians is cause for consumer concern. The NRC urges policy makers to help recruit veterinarians. They suggest that experienced veterinarians will be essential to address increasing demands for meat throughout the world.[3] Animal disease experts will also play increasingly important roles as they help monitor and manage animal epidemics in our globalized food supply.

These challenges for rural veterinarians and animal farms come at a time when animal health and food issues are growing public concerns. Consumers today expect American animal farms to exercise production practices that are fair, sustainable, and healthful while simultaneously producing affordable products. Questionable practices limit our success in meeting such goals. These practices have been brought to consumers' attention in a variety of ways. *King Corn*, *Weight of the Nation*, and *Genetic Roulette* are only a handful of popular food-centric movies that highlight contentious practices on modern factory farms. The producers of these films are joined by an array of well-known authors and food advocates hoping to further various food related agendas.

Yet, despite growing concerns, few public policies keep remaining animal farmers in business. As a result, we live in a nation that is importing increasing amounts of meat and dairy products while concurrently relying on less than 1% of its own population to oversee farmed animals.[4] Farmers who have traditionally enjoyed treating each animal with the attention and concern it deserves struggle to compete within our national marketplace.

Sadly, animal agriculture may go the way of America's clothing and textile industries. As a nation, we can either continue consolidating and shipping food production overseas, where practices are less visible and accountable to the American people, or we can work harder on what we have left: paving the way toward a less concentrated, productivity-oriented agriculture, toward practices that are more mindful of regional values, jobs, natural resources, and health.

Doing so will require policy and regulation changes that encourage on-farm diversity and animal health. We already rely on too few farmers and too many giant monocultures as it is.

Some consumer groups are demonstrating that Americans can create enough market pressure to positively influence farm practices. In the last decade alone, Americans have watched the demand for farmer's markets, community supported agriculture, and fresh foods grow substantially. Some young adults are even forgoing more economically promising careers to work farmland, actively participating in movements to bring healthy, local foods to American tables.

Still, states with agriculture-oriented economies resist many of these changes. The healthy, local foods prevalent in areas along the coasts and in college towns are not nearly as visible in rural regions of states like Iowa, South Dakota, Missouri, and Indiana. Rural communities that need new small and medium sized farms the most are seeing little of the American food revival.

Proposed changes to ag industry structure are often perceived as too great of a risk in these areas. Those who have remained in animal agriculture have often already increased production substantially or sacrificed a great deal to remain on the farm. Remaining small farmers in my father's practice typically rely on large loans and multiple sources of income to supplement their farm activities.[8] Here consolidation continues to take its toll on jobs, health, and ecological diversity, especially in the most impoverished regions.

In the last ten years, I have had multiple opportunities to observe regional differences in acceptable food production practices between my home state and an area of western Massachusetts called the Pioneer Valley, where I now live and teach. During this time, I have been especially impressed by the number of Americans interested in supporting small to medium sized, value-added farms in the Pioneer Valley region. In this area, more ecological farms are successfully feeding tens of thousands of eager customers every year.

It strikes me as odd that this region, which has far less arable land and a much greater population density than Iowa, should be having so much more success on this front. I attribute most of this success to different socioeconomic and cultural factors. On average, people living in the Pioneer Valley have more disposable income to spend on food products. They also have a higher level of skepticism toward mass production practices than the typical Iowan. Numerous colleges and restaurants help to absorb the extra costs associated with value-added farms scattered throughout this food-centric region.

In addition to these differences, producers in the Pioneer Valley don't have to worry about the possibility that agribusiness production practices might interfere with their endeavors on the farm.

Last year, before returning to Iowa for a summer visit, I told my husband that I couldn't wait to be back in Iowa, free from the incessant hum of east coast air traffic. While I have grown accustomed to many aspects of the Pioneer Valley, I will forever be homesick for the calm I associate with rural Iowa. People in rural Iowa aren't racing to and from work on congested interstates or subject to the sensory overload that comes with Massachusetts's relatively high population densities.

After a few days in Iowa, I had to eat my words. We'd returned to an agricultural battleground: low flying planes dumping chemical herbicides and fungicides on thousands of acres of genetically modified crops below. The peaceful outdoor experiences I associate with my rural upbringing were constantly interrupted by crop dusters—planes hired to spread chemicals across these fields. I couldn't help wondering when crop dusting had become such a necessary and acceptable part of the Iowan landscape and culture.

Organic farms, like my brother's family farm, have to work within this heavily modified ecosystem. In areas where crop dusting is common, fence lines offer little protection for the few farmers attempting to return to organic production practices. On more than one occasion, my brother's family has had to take measures to ensure that other farmers' planes don't accidentally void his organic certification status while distributing chemicals near his property.

In addition to these challenges, value-added farmers in ag oriented states must be willing to drive many more miles to find a willing customer base. Unfortunately, many rural Iowans don't feel they can afford to purchase the value-added organic and whole foods associated with sustainable production practices. Such views discourage attempts to transition toward more sustainable farm models because these products are perceived as being too expensive for the average worker.

Transitions toward healthier, more diverse farms are further limited by misinformed ag policies that encourage ag monocultures. In the following paragraphs I will explore a few of these policies before moving on to discuss how they have misshaped public perception of important animal health issues.

One government incentive that has negatively impacted animal agriculture is the Renewable Fuels Standard, better known in many Midwestern states as the "Ethanol Mandate." Historically the RFS has required that certain volumes of biofuels be blended into gasoline products before being sold at the pump. During the 2012 growing season, this mandate drove approximately 40% of American corn into ethanol production. Our government offered ethanol producers a tax credit of 45 cents per gallon to produce renewable corn fuel, a subsidy that cost taxpayers billions. The RFS essentially guarantees continued demand for a crop system that is now well established.

High demand for corn puts pressure on our nation's animal farmers. Contrary to the widespread notion that most of the nation's "breadbasket" is brimming with food for the masses, a large percentage of American corn is actually used to fuel cars and trucks. This demand for ethanol creates a system that encourages competition between the fuel sector and food supply, because corn is a main component in animal feed.[5,6,7]

When livestock farmers took an especially hard-hit during 2012's record-breaking drought, high feed prices threatened many animal farmers' livelihoods. While corn farmers could rely on the RFS and high-quality crop insurance, livestock farmers had relatively little assistance.[7] They watched nervously as feed prices soared, wondering if they would be able to stay in business while keeping their animals properly fed and watered. Stressful situations like these have created a growing pride-based tension between many remaining animal farmers and farmers whose sole industry is commodity crop production.

Livestock farmers receive a mere 12 cents to the dollar for the animals they sell. Most of the retail price for animal products goes to giant food processors and retail stores. In addition to being paid less per pound of meat produced, farmers have to deal with a customer base that puts less value on their work than the value assigned to similar products in other countries. Americans spend approximately 7% of their annual budgets on food, versus 15% in Japan, 21 % in China, and 41% in Kenya.[7]

Other policies only add to these tensions. Even though congress is planning to reduce crop subsidies, it has continued to approve a disaster-relief insurance program to supplement the existing federal crop insurance program. On top of disaster support, these programs offer corn and soybean farmers premium subsidies that protect them from common market fluctuations.[8] These insurances further distort food markets in addition to the litany of federal handouts crop farmers have already received in the last several decades.

Sadly, most of this federally supported aide has been awarded to America's largest corn and soybean farmers, many of whom have received more than one million dollars per year in crop insurance subsidies! To make matters worse, in the last ten years insurance companies have collected roughly half of the $50 billion spent supplementing crop policies. American taxpayers assume this money goes to farmers.[8]

These federal funds are not tied to improved conservation practices, on-farm diversity, or rural health. In fact, as part of proposed budget cuts, farm conservation programs are often put on the chopping block. This provides further incentive for farmers to buy profit-guaranteed policies for plantings on steep, rugged terrain that shouldn't be planted to corn and beans in the first place.

No amount of political or industrial support for corn-ethanol will change the fact that it is less fuel-efficient than regular gasoline. Still, government supports like the RFS could be a good idea if they were tied to other renewable fuel developments. Instead, past federal packages have been directed toward corn. These packages provide relatively little help to animal farmers, those hoping to diversify their crop bases, and those raising non-insured produce.

While debates over misinformed policies like the RFS and federal crop insurances highlight growing tensions between animal farms and crop farms, public conversations about farm management practices often only exacerbate them. Unlike Iowa's animal farmers, corn and soybean farmers don't sit in the middle of heated and well-publicized debates regarding animal welfare. These debates occasionally result in decisions that actually decrease overall welfare because they are applied in a context many consumers don't understand. Some of these conversations have only encouraged remaining animal farmers to sign contracts or give up meat production to pursue less stressful, more economically rewarding opportunities.

During this time of economic turmoil for low-income and jobless Americans, it seems particularly important to reevaluate rural America's pitfalls. Main streets throughout rural America would gladly welcome increased demand for labor and intellectual resources, especially if these resources are properly valued. The agribusiness publications that boast job creation seem almost comical to my father and I as we drive through the towns of Hopkinton, Ryan, and Delhi, Iowa. Residents here don't need marketed statistics to know that this data doesn't tell the whole story. All they have to do is open their eyes to the lack of job opportunities around them.

Like many other small towns in the Midwest, in the last two decades Hopkinton has cut or reduced hours at its bank, grocery store, post office, and library as well as several other independently owned shops and businesses. If you want to work in Hopkinton now, one of your only options is to work for a company on the outskirts of town called IAS, Innovative Ag Services, a company that focuses its services on three markets: agronomy, feed, and grain. *Innovative* ag services include consultation, analysis, and sales of fertilizers, pesticides, genetically modified seeds, and premixed livestock feeds. An ag specialist at IAS can help your farm figure out which fertilizers, or *plant foods*, and chemicals, or *crop protectors,* as they are called here, are appropriate to optimize crop production.

If you are qualified to operate large machinery, you can apply to be a crop production specialist or chemical applicator at one of IAS's small town locations. You can also work in sales, grain operations, or transportation.[9] Workers can make a decent living with benefits in an area where such employment opportunities are hard to find. Consumers across the nation, demanding access to cheap, highly processed foods keep these chemical crop protector, fertilizer, and GM seed services in high demand.

The genetically engineered seeds and chemicals sold at places like Innovative Ag Services help American farmers farm more acres. Agribusiness proponents also argue that these technologically advanced products have helped customers achieve yields that are 250% greater per farmer. They report that the average American farmer can produce food for 155 people today compared to only 10 people in 1930.[10]

Quite frankly, when I look at my home state, I'm very surprised that these figures aren't larger. The reported statistics leave out the part about our nation's reliance on far fewer farms total: approximately 25% of American families still lived on the farm in 1930 compared to less than 2% today, as well as the fact that most industrial farms are much larger than they were then. These farms *should* produce 250% more *human* food than they did years ago. Instead, many yield comparisons also overlook the fact that less than 5% of the high-yield corn raised in America actually provides nutritious food for American families.[11] The majority is used to make ethanol, animal feed products, and high-fructose corn syrup.

Some of our cheap corn is then dumped overseas in the name of foreign aide. Agribusinesses like Monsanto, Dupont, and Syngenta would like American farmers to believe that these increased yields reach half way around the globe and land on the plate of a poor, starving African child. This, despite the fact that an African child may be better off without America's cheap, genetically engineered seed products. Despite the fact that foreign farmers can raise their own, more nutritious crops in toxin free environments. Despite the fact that these same foreigners might one day like jobs farming native crops that have not been genetically altered by companies they don't own.

These mixed messages are regularly on display in several of the most popular farm publications. Among these magazines, my father reads *FeedStuffs: The Weekly Newspaper for Agribusiness*, the *Hoard's Dairyman*, the *National Pork Producer*, and various *Checkoff Reports*. The articles and ads in these publications encourage farmers to buy products and request medications that will boost productivity to help feed the world. By reading such publications, veterinarians can better understand their clients' requests and concerns. The articles provide an informal education of sorts, and, by many, are read religiously.

Thanks to the wide and incessant dispersal of marketed perceptions, including the perception that we must increase food productivity at all costs, many farmers genuinely believe that their thousand acre monocultures, GM seeds, and routine chemical applications are, without doubt, saving the environment, feeding the world, and keeping animals healthier than they otherwise would be.

While giant agribusinesses campaign to feed the world, spending large amounts of money to convince farmers that some difficult, and arguably unethical, production choices are for the greater good, grassroots organizations, many of which include former farmers, are working together to deliver a different message. They tell us that intensely diversified fruit, vegetable, and animal farms already have the capacity to feed many more *people* per acre than crop monocultures do now.

In the longest side-by-side comparison of conventional farming to organic farming, the Rodale Institute came to the conclusion that organic yields can match conventional yields in long-term trials and outperform conventional yields in years of drought. The Rodale Institute also reports that organic systems build rather than deplete the soil, use significantly less energy, produce far fewer greenhouse gases, and are just as profitable as conventional systems in the long run.[12]

In a similar study led by Iowa State University's Long-Term Agroecology Experiment, researchers compared an organic soy-corn rotation to a conventional soy-corn rotation. They found that average yields for organic corn, soybeans, and oats were equivalent to or greater than their conventional counterparts. Because these organic systems required fewer inputs, average returns after deducting labor and production costs, were also greater. Many of these gains were made without genetic technologies and wastes.[13] Recent yield comparisons provide further evidence that progress in conventional yield improvements has slowed in the last decade. Unfortunately, these high yielding crops are often yielding less widely reported damage to our health and environments.

Given that more than one-third of adults and 17% of children and adolescents are overweight, it seems downright deceptive that we should allow American companies to imply that their profit-hungry missions to feed the world create a healthier planet.[14] This figure does not include the millions of Americans suffering from mental disorders, including depression and ADHD, that many reputable scientists now link to an unhealthy diet and environment. If our hearts are in the right place, our heads are torn between two camps: some Americans still adamantly attempting to export the notion that foreigners "need" an Americanized food supply while others try to convince foreigners that they don't want what we have. It's time to revisit our mission to feed the world with a more critical eye.

Unfortunately, these mixed messages leave American farmers and consumers in a tight spot. Proposed farming methods that rely more on human labor and intellectual resources, and less on huge amounts of fuel, water, chemicals, and bioengineered seeds, should be encouraged through American farm policies. Agribusiness messages that pit citizens who are legitimately concerned about animal health, human health, and the environment against progress, productivity, and marketed goals should be discouraged. We cannot be so divided if we hope to address current food production challenges, including the challenge to keep remaining animal agriculture jobs in the United States.

In the last two years, I've come to appreciate just how strong this sense of dividedness is. Even when disparate parties have common interests, industry leaders use their publications to mislead the public. The very language we use to address our ag problems has become two-headed: our words meaning one thing in one context and entirely another elsewhere. Within this context, intelligent scientists are afraid to voice their concerns because they have been perpetually bombarded by publications that deem such thoughts kooky and extreme. My dad and many of his veterinary peers occasionally fall into this population. They know that most of their clients are reading agribusiness publications and sometimes fail to escape the influence of two-headed perceptions themselves.

This attempt to divide and manipulate language is perhaps most evident in a variety of federal promotion publications called *Checkoff Reports* that are delivered to American farmers on a regular basis. These reports are published using *Checkoff* funds—funds created to generate generic demand for products like beef, pork, dairy, and some fruits and vegetables. The advertisements supported by this type of fund promote a type of food product rather than a particular brand. *Got Milk?* and *Beef: It's What's For Dinner* are examples of generic slogans most Americans recognize today.[15,16]

To fund a checkoff program, food producers and handlers pay an assessment tax, usually at the time of sale, called a *checkoff fee*. The *Beef Checkoff* collects $1 per animal sold to fund beef advertising, industry research, and education. Checkoff supporters argue that this helps beef producers expand their markets. Opponents contend that a portion of their sale funds advertisements and research they wouldn't normally pay for.[15]

While voluntary checkoffs do exist, mandatory contributions to checkoff programs have become increasingly common in the last three decades. Some farmers argue that mandated checkoff messages are a violation of their first amendment rights. The Supreme Court has historically upheld that this argument is not enough reason to overrule the majority of farmers and politicians who feel the benefits of such programs outweigh the costs.[16]

As consumers become familiar with modern industrial ag practices and voice concerns that ag practices should become more sustainable and transparent, the majority of farmers and politicians may begin to change their minds. This is unless, of course, current checkoff reports can convince consumers that such changes have already occurred. In the ensuing section, I explore how current checkoffs twist language to give farmers and consumers this false impression.

In a recent article in the *Pork Checkoff*, the President of Communications, Michael Wegner, gives pork producers tips on how to address concerned consumers. In his article, "Lost in Translation," he reports that *eye-opening* research conducted by the Farmers and Ranchers Alliance (USFRA), reveals that "one of agriculture's main messages – that we produce safe, affordable and abundant food – doesn't work any longer" and has inspired food industry language experts to push "new communication strategies." Wegner encourages his readers to adhere to several new "pieces of wisdom" thanks to this research.[17]

Unfortunately, much of Wegner's advice sounds like a parent encouraging deceptive adolescent behavior. He encourages his readers to "use language that reminds people of the wholesome elements of what you do," when discussing farm management methods. "Instead of talking about *using pesticides*," he urges crop farmers to talk about "preventing bugs and other pests from eating crops." Instead of *using fertilizer*, crop farmers should say that they are *nurturing crops with the nutrients they need to grow*. Instead of planting *genetically modified crops*, these farmers should continue using *seeds developed to grow stronger, more resilient crops*![17] Instead of doing your own homework, consider *effectively utilizing your resources* to *obtain* Sally's homework instead. Then put your name on the top, and hand it in as your own.

This twisted rhetoric is very similar to the wording used on the Innovative Ag Services website to market fertilizers, pesticides, and seeds. Wegner is right: our *words matter*. Unfortunately, honesty doesn't seem to matter as much. In this very same article, Wegner points out that only 20 percent of people give agriculture an eight, nine or ten when asked to rate their trust in the industry.[17] He seems surprised! It's no wonder we have a difficult time trusting our industrial ag leaders!

The use of two-headed language is typical throughout the entire checkoff publication. In a later article called "In Defense of Choice," Chris Novak, the CEO of the National Pork Board, writes in response to McDonald's recent announcement that it will phase out purchases of pork supplied by factory farms using gestation crates. During the last two decades, 60-70% of U.S. breeding sows have been confined to crates throughout their 112-115 day pregnancies, many in spaces so small sows can't even turn around.[19]

Without mentioning the numerous health ramifications of such practices, Novak pits animal welfare groups against farmers. He argues that the McDonald's decision to back customers' requests that crates be phased out, "took something away from you [farmers] – your ability to make the best decision for your animals with the advice of your veterinarian and other swine care professionals...The science supporting gestation stalls is clear – individual housing of sows protects them from aggression found in larger groups and allows individualized care."[18] Novak fails to mention that research has also linked this aggression to tight quarters, and that many pork producers already lost their choice in the matter years earlier when they signed corporate contracts with giant pork monopolies.

Compare these words, with those in the Humane Society's report on the same topic, a publication that cited numerous peer reviewed animal behavior, vet medicine, and foreign government articles. They found that, by and large, the science supporting gestation stalls proved just the opposite. In fact, scientific researchers found that crated sows have suffered a number of problems including increased risk of urinary infections, weakened bones, overgrown hooves, and lameness.[19]

While my father is typically skeptical of Humane Society reports, because each comes with its own set of biases, he has observed too many of these negative consequences firsthand. While some farm operators are more attentive to animal welfare issues than others, it is very difficult for any farm operator to prioritize animal health when an animal lives in a tiny crate. Though several of his clients have learned to manage crated sows to the best of their abilities, few seem particularly proud that they need to use crates to stay competitive. Nonetheless, crates have become the industrial norm. In this case, our innovations and technological advances have done little to advance human concern for basic animal welfare and health on the majority of modern sow operations.

Sows are not the only animals that have been enclosed in tight spaces for considerable durations. During the last three decades, high-density confinement operations have used stalls to separate dairy calves and laying hens as well. The densities and dimensions of these enclosures restrict movement and socialization with other animals but are provided because they allow the farm operator more economic wiggle room in a very competitive market. Using such enclosures, the confinement operator can raise more animals in one space while reducing the risk of disease transmission and aggression between tightly packed animals. Enclosures can also improve workers' abilities to feed and monitor each animal when animals are raised at such densities.

Confining any animal to indoor enclosures doesn't come without health costs. In addition to the welfare concerns that go unaddressed, my father points out that animals living in confinement often experience a wide range of symptoms associated with low levels of vitamin D and exercise.

Not surprisingly, animals, like humans, suffer from vitamin D deficiencies when they do not obtain enough sunlight. The vitamin D produced in the skin when it is exposed to UV radiation is essential for healthy absorption of calcium and phosphorus in the intestine. Like humans, animals with low vitamin D can develop rickets, a condition in which the bones become weak, brittle, and sometimes fracture. While most operators are mindful of these conditions and purchase vitamin D supplements to prevent animal bone diseases, my father has also been to operations where too many animals have weak, brittle, and fractured bones. Many of these conditions could be easily remedied with a mere 30 minutes of sunlight per day. He's found that synthetic vitamin D supplements are not always as reliable as purported.

In addition to its effect on bone quality, Vitamin D affects the immune system. In fact, some scientists associate serious winter flu outbreaks with decreased sun exposure. When individuals are confined indoors, they experience reduced immune responses.

Researchers at the National Animal Disease Center think that vitamin D may have a similar effect on cows. Vitamin D can actually reduce the severity and prevalence of mastitis infections in dairy cattle, a painful infection in the cow's udder. In one study, researchers found that cows receiving udder injections of natural vitamin D had far lower bacterial cell counts and infection rates compared to other cows. Milk production was also greater in treated animals.[20]

This research may have implications for human health as we try to avoid growing dependencies on antibiotics. If veterinarians can treat mastitis with Vitamin D or prevent issues like mastitis from occurring, they can rely on fewer antibiotics overall. At present, mastitis is one of the most common and costly diseases of American dairy cattle.

Humans can also suffer from mastitis if they choose to nurse their babies. I should know: I had mastitis twice after giving birth to my son. If the bacterial infection is severe enough, a doctor, like a veterinarian, will prescribe antibiotics. Interestingly, after being prescribed antibiotics, I later discovered that I, myself, had a Vitamin D deficiency. Perhaps a few more postpartum walks in the sun would have served me well?

Walks have numerous benefits of their own. Not surprisingly, animals who get frequent exercise tend to be healthier and more productive.[21]Sadly, some of the chickens and pigs my father has treated in confinement are so weak, from a lack of vitamin D and an inability to exercise, they cannot walk. Unfortunately, for the animal, difficulties walking do not necessarily reduce a pig's economic value in the marketplace unless it can't walk off the truck at sale. If a pig goes down at the slaughterhouse, the farmer typically has to pay for the pig to be hauled away.

While prominent veterinary groups in the United States have not openly condemned confinement practices, the European Union Scientific Veterinary Committee (SVC) hasn't been as tight-lipped with its criticisms. More than a decade ago, the SVC concluded that individual pens that do not allow sows to turn around easily should not be used.[19] According to the Humane Society report, even though there was a "clear international trend" toward gestation crate reduction, gestation crate use in the United States remained "a common animal agribusiness practice."[19] Have American veterinarians been so worried about being associated with animal welfare activists like the Humane Society that they failed to speak their minds? Even when the evidence overwhelmingly indicated that gestation crates and confinement units have not been healthy for American livestock?

In these types of conversations, American veterinarians sit in a uniquely complex position: their job is to provide care for animals destined for the plate. Yet, many, like my father, were called to the profession because they enjoy working with animals and genuinely care about their health. In fact, all American veterinarians, like American doctors, have sworn an oath:

> Being admitted to the profession of veterinary medicine, I solemnly swear to use my scientific knowledge and skills for the benefit of society through the protection of animal health and welfare, the prevention and relief of animal suffering, the conservation of animal resources, the promotion of public health, and the advancement of medical knowledge.
>
> I will practice my profession conscientiously, with dignity, and in keeping with the principles of veterinary medical ethics. I accept as a lifelong obligation the continual improvement of my professional knowledge and competence.[22]

Unfortunately, veterinarians' abilities to protect animal health and welfare have increasingly come into conflict with productivity and efficiency measures.[23] Too many of these goals have inevitably gone by the wayside.

Veterinarians who have been around for thirty years, like my father, cannot deny that animal health and welfare have been seriously compromised as American farmers have strived to meet ever more demanding bottom lines. These veterinarians aren't naïve, or intentionally ignoring their principles, but they are in a very tricky position. If they want to stay employed, they have to work in buildings where animal welfare and health are no longer the priorities they once were. Novak doesn't think twice about the parameters of his conversation, because he doesn't question the difficult decisions many farmers made to remain in business. Outside this occupational sphere, consumers press farmers and veterinarians for better explanations. They reasonably wonder why a pregnant sow, even if she's bound for the plate, ever needs to sit in one spot for an entire four months. Does the quality of an animal's life mean nothing?

Novak does make one point the Humane Society largely misses: hog farmers struggling to keep up with the biggest-is-best mentality will have a difficult time transitioning back to non-crated systems. In addition to signing corporate contracts, many industrial pork producers have signed up for life long debt: constantly needing to reinvest in items required by their corporate contract managers. For the industry's largest producers, transitioning to non-crated systems might cost hundreds of thousands of dollars. It very well could be the straw that breaks the camel's back.

And when pigs are released from their crates, management may become even more difficult in modern facilities. Farmers will need to decrease production densities to quell aggressive pigs kept in tight but open quarters. They will have trouble trying to house sows together in these huge 1200 plus sow operations where most pigs today are farrowed. Is a crate the lesser of two evils when profit is so density dependent? Farmers and veterinarians working in high-density operations may understand the reality of the situation much more than they like to admit. If operators continue to expect the same profit margins from these operations, a crate is just as much a refuge as a prison.

One must be able to think outside the box to make a truly ethical decision. While many farmers are listening and trying to respond to consumer concerns, animal farmers will need a great deal of public support to transition away from conventional confinement practices toward less dense environments. Farmers and consumers need to work together to ensure that alternative settings truly address animal health issues while being mindful of the demands placed on transitioning farms.

In any case, it seems appropriate that we stop thinking of animal welfare groups, veterinarians, and farmers as oppositional and start thinking about ways we can collectively raise our expectations. Quality independent farmers and veterinarians are in great positions to be excellent animal caretakers, but if they have no place to work because the demand for cheap product drives animal production elsewhere, quality caretaking skills won't matter.

When consumers and industry leaders work together to overcome obstacles we can create positive change. In the aforementioned case, consumers got on board with one of the world's largest food corporations, McDonald's, and created demand for a market-driven production change. Several other fast-food chains and grocery stores followed suit. As a result, fewer pigs are being confined to crates for months at a time. Now these same producers and consumers need to make sure such changes don't come with unintended consequences. Our words need to clearly represent our work.

While the *Pork Checkoff Report* assures farmers that the pork industry is making ethical decisions using the *best science available,* it isn't clear what they mean by this. Have the authors spoken to farmers and veterinarians who *have been* concerned for years about their pigs, too many of whom die unnecessarily in confinement? Throughout the report, authors imply that their goals are to improve animal *welfare* and *sustainability*. Clearly, these words have been completely corrupted. One author asserts that the facts in his article are clear and that the U.S. Pork Industry has seen "decades of continued improvement and sustainability." He goes as far as to quote John Adams saying, "facts are stubborn things."[24]

Such facts describing a more sustainable industry are completely dependent on previously defined parameters. In this context, present inputs are more efficient than similar industrial practices a decade ago but are not compared to farms a century ago or more diverse systems today. The weight-gain and animal-use efficiencies used to justify industry's smaller, greener footprint ignore larger inefficiencies accumulating in non-recycable wastes, fertilizers, manufacturer costs, transportation, buildings, air filtration, lagoon systems, nutrient supplements, veterinary medicines, and energy.

Any good farmer who raised pigs before signing a contract or building a confinement unit knows that this data is a huge oversimplification. On top of this, these measurements do not include any job or nutrient quality data—much of the pork that is raised by far fewer people contains questionable health benefits for both hogs and humans.

We can't ignore any input or output in our conversations about health and sustainability. This is easy to see when comparing animal waste disposal systems. In CAFOs, the liquid manure becomes a source of pollution as toxic compounds accumulate over giant manure pits. The industry, however, reports that we are getting better at managing this waste because we are finding new ways to reduce and contain these toxins. Hogs rotated through pastures, on the other hand, spread their manure around for free. These free wastes fertilize the soils and microorganisms living there. This poop adds valuable nutrients and minerals back into the soil while allowing animals a chance to exercise. At the same time, the recycled poop doesn't have the chance to accumulate in one area becoming an environmental hazard. Though *Checkoffs* can justifiably boast that we have *improved* some efficiencies, we have also created inefficiencies we never needed to address in the first place!

Because this outdoor pig requires more body fat to stay warm, it can also retain more fat-soluble vitamins that are beneficial to both pigs and human pig-eaters. Fat-soluble vitamins include vitamins A, D, and E. Confinement hogs, on the other hand, have been bred to produce leaner cuts of meat. Many could no longer survive outside in an Iowan winter if they tried. They simply don't have as many essential fat reserves or the vitamins they hold.

In the absence of transparent parameters, industrial ag reports often seem to assume that I, the reader, only value one type of efficiency: economic efficiency. I am supposed to be glad that industry has come up with a means to use fewer pigs to produce the same amount of pork, even though I'm aware of associated animal welfare and health costs. I'm supposed to be happy that it requires way less land to raise pork, even though I know that the quality of a pig's life is seriously compromised. Furthermore, I am supposed to be happy that the number of farmers employed to raise the same amount of meat is less, even though there is more unemployment and less care given to each animal. I can take all of the money this cheap pork is saving me and spend it on rising health care costs and disaster relief associated with my super cheap diet. This article seriously underestimates both my intelligence and my ability to value things other than the American dollar.

These publications also put our nations' animal caretakers and land stewards in positions where they feel pressured to defend such activities—activities that will never be publicly acceptable because they are widely viewed as wrong—and rightfully so. Instead of funding research designed to justify indifference and harm, we should encourage farmers to tackle their animal health and welfare concerns head-on.

Using science to sway public perception rather than changing such practices makes our nations' food producers and businesses appear morally inept. These producers seem incapable of making positive decisions themselves. As my father has observed time and again, this is typically not the case. In fact, on many of his daily farm visits, clients must make difficult, but positive decisions to save their sick and dying animals. Still, the public is left to doubt: who, exactly defines what is right? Is it the misinformed media spin-machine, the misinformed farmer reading spin-machine publications, or the consumer who, no matter the number of statistics tossed his way, still doesn't feel right about farms where sows are caged for the entire duration of their pregnancies?

As a teacher, I frequently remind my students that one of the first things we teach American grade-schoolers is to tell the truth. Since when is it okay to be so obviously deceptive? Some education gurus suggest that teachers aren't adequately preparing students for the real world. Perhaps we're teaching kids to be too honest in a system that has absolutely no respect for honesty, our children's hearts unable to handle the pressure when their bosses want them to lie about their work?

Just as my father couldn't save our mutant two-headed pig, he has a difficult time saving some of the creatures feeding on, and living in, the genetically modified, heavily treated world of agribusiness in which he now works. In the past, such creatures were the exception rather than the norm. Today all kinds of unusual plant and animal health problems arise in these fields, including a greater number of two-headed social, political, and environmental problems. Though industry reminds us that it is hard to tie health problems to a specific chemical or genetic input, my father and I are attempting to prove that the evidence against several toxic chemicals, practices, and policies is already overwhelming.

DIAGNOSIS 2: MANGANESE DEFICIENCY

A good doctor's mind is like a giant funnel. When addressing symptoms, he or she starts at the top, considering a wide range of possible causes before narrowing the scope of his or her search. A good doctor understands that disease often develops from many factors, not just one. Using the process of elimination, the doctor hopes to arrive at a diagnosis and treatment.

Disease, too, can be funnel shaped. We observe a wide variety of symptoms that stem from some cause. Treatments and diagnoses are simplest when the bottom of the funnel narrows to a single source. Complications may arise as the number of sources increase. Such is the case with many of today's industrial-ag induced ailments.

When my father initially attempted to identify the source of manganese deficiency in several cow herds, he found it difficult to get to the narrow end of the funnel. The symptoms were classic, among them the cows had pea-sized ovaries, irregular reproductive cycles, reduced conception rates, and generally low productivity. Severe manganese deficiencies can result in such badly deformed fetuses that the cows miscarry or deliver stillborn calves. When the manganese-deficient cows do bear live young, the calves are often weak and small. Some suffer from enlarged joints, limb deformities, dwarfism, and shortened facial bones. Unfortunately, the source of the deficiency in my father's practice remained elusive.

His interest in solving this puzzle increased when he realized that manganese deficiency had become more prevalent in his practice than he initially realized. Before 2000, he had not seen a single case of it. Imagine his surprise then, when several herds he'd been working with for decades began to exhibit telltale signs of manganese deficiency. In one beef client's herd, the problem was so bad that 50% of his calves exhibited skeletal deformities, including disproportionate dwarfism. What had changed in the last twenty years to warrant an increase in cases?

Answering this question proved difficult from the start. Diagnosing manganese deficiency is not a straightforward process. Simple blood tests can be misleading because red blood cells often contain higher amounts of manganese than surrounding blood serum. If these blood cells rupture during transportation, manganese leaches into the surrounding serum and inflates overall test results.

Blood tests are also a wash if manganese is present but unavailable to the animal. When the amount of manganese oxide is higher than manganese sulfate, the cow may be unable to absorb enough manganese even when it's detectable in the blood stream.[1] A cow then might struggle with manganese deficiency even when its test results are normal.

For these reasons, the best way to confirm manganese problems is to send a sample of a dead animal's liver to a lab. This is much more reliable than blood work alone because the liver is an important storage organ and can more accurately show when an animal has metabolized this nutrient.

However, if none of the cows have died, manganese supplementation may be necessary to confirm one's diagnosis. If you feed the cows more manganese sulfate and they get better, you know what the problem is. The disadvantage to this method is that you have to wait to see if it works – a process that can take several months. If the problem is affecting 50% of your herd and you're watching your cows give birth to stillborn, deformed calves, waiting for months may not seem like an option.

In the cases confronting my father, several animals' liver samples came back positive for manganese deficiencies. This left little room for ambiguity. In fact, wet chemical analysis of livers from one herd in 2006 could not detect manganese at all! The lab analyst told my father that levels below the detection limit were unheard of for clients feeding their cows Iowan corn, soybean meal, and hay—all feed items that should contain enough manganese to show up in a liver analysis when animals are eating an otherwise balanced diet.

Needless to say, my father was intrigued. Why were the manganese levels so low in diets that had historically provided enough of this key nutrient? After doing some research, he decided to test more livers. In addition to confirming several other cases in dairy and beef cattle, he later confirmed that hogs owned by clients who had experienced high hog mortality rates due to the common hog viruses, PRRS and PCV, were also manganese deficient. Were his clients' hogs dying more frequently because manganese deficiency increased their susceptibilities to disease?

He got an important clue in 2007 when a client showed him an article in the "John Deere Furrow." According to the author, spraying manganese on soybeans 10-14 days after spraying them with glyphosate, the active ingredient in the herbicide Roundup, could significantly increase plant yields. The article was written by a Dr. Don Huber, an experienced plant pathologist.

Huber argued that these herbicide formulations were causing some crops to become manganese deficient and susceptible to disease. The glyphosate sprayed on genetically engineered crop varieties, as well as the glyphosate used for pre-plant weed control, was binding to nutrients in the soil and plants. Sprayed crops were less able to metabolize the bound nutrients needed for proper plant function. As with my father's animal tissue samples, special chemical analysis was conducted on plant tissues to determine that these nutrients weren't available to the crops. Certain soil types, specifically the sandy, high pH soils common in many parts of northeastern Iowa, seemed more prone to glyphosate induced problems than others. By foliar fertilizing these deficient plants with manganese and zinc, Huber argued that farmers could increase bean yields by as much as 8-12 bushels/acre.

Huber's concerns about plants were very similar to those my dad harbored toward diseased pigs and cattle. Dad promptly phoned Dr. Huber and, after extensive conversation, began to read his book, *Mineral Nutrition and Plant Disease*. Reading Huber's book put him one step closer to uncovering the sources of his clients' ailments: it seemed as if it might be a problem with the animals' diets.

In addition to researching such similarities, Dad kept his eyes and ears peeled for other news related to manganese deficiency. At a Minnesota Dairy Health Conference, he listened to Dr. Jeremy Schefers' lecture on cattle stillbirths associated with manganese deficiency. Drawing on lab tests performed on sixty-two Midwestern herds, Dr. Schefers found that calves with skeletal deformities did not have enough bioavailable manganese in their diets.[2] Like Huber, he suggested supplementation—in this case adding manganese to pregnant cattle's diets at a higher level than the 2001 National Research Council's recommendations.

Dr. Schefers also acknowledged that the relationship between glyphosate and manganese deficiency had gained attention among livestock producers and agronomists. Yet, he wasn't sure if glyphosate could cause so many herd deficiencies. Schefers pointed out that manganese deficiencies in plants are not uncommon where soil has a high pH and low soil moisture. He also suggested that the observed manganese deficiencies could have other environmental sources associated with these conditions. While much of his talk confirmed my father's suspicions, his suggestions also created new questions.

While reflecting on Schefers' presentation and a variety of articles on manganese deficiency, my father came to the conclusion that he could rule out the environmental factors proposed by Schefers. While manganese deficiencies are nothing new to the planet and can be associated with precisely those soil conditions described, he was seeing an unprecedented number of cases in his decades-long practice. Furthermore, the period during which he documented an unusual amount of manganese deficiency was not a period of low soil moisture, as was suggested, but rather very wet. In fact, it was so wet that the nearby Lake Delhi damn burst. Because cases were popping up in areas that had previously seen no cases of manganese deficiency, high soil pH also seemed an unlikely contributor. For these reasons, he returned his attention to the way feed crops were being raised.

From farm to farm, crop pesticides, like glyphosate, and the genetically modified (GM) crops engineered to withstand them, were feed variables that had consistently changed. As he reflected upon thirty years of practice, several relatively new dietary illnesses, and an abundance of scholarly articles linking herbicides to animal ailments, he became more convinced that he had a possible answer: chemical applications were somehow reducing the amount of bioavailable nutrients in the crops used to feed hogs and cattle. He convinced several clients to give their livestock feedstuffs that hadn't been sprayed with glyphosate-based herbicides for weed control. When the animals' health improved, he knew he was onto something.

His evidence is not bulletproof. Unfortunately, research comparing the long-term effects of glyphosate-based herbicides on GM crops to non-GM crops is lacking. When diagnosing today's feed-related nutrient deficiencies, veterinarians often have to find evidence in a rather backward trial-and-error fashion. In my father's experiences with manganese deficiency, he had to rely on clinical observations and experimental feeding regimens to correct herd health problems.

But this is our food-supply we're talking about. Many of us do, ultimately, eat the cows and pigs being fed herbicide-treated, genetically modified corn and soybeans, as well as products containing these same crops. If the animals are unhealthy, how healthy are these

glyphosate disrupts production of these proteins weakening the plant's overall immune response. Glyphosate ultimately damages or kills any species that depends on the naturally occurring EPSPS enzyme – that's why it's such a good pesticide. Because so many organisms utilize this enzyme, including all green plants and an unknown quantity of microorganisms and fungi, one application of glyphosate can impair many soil-bound species.

Scientists created glyphosate resistant crops after identifying a strain of bacteria called CP4.[3] The strain was isolated from a waste site at a herbicide production facility. While many bacteria cannot survive in glyphosate mediums, CP4 grew uninhibited amidst glyphosate residues. This is because CP4's EPSPS enzyme has unique structural properties for a shikimic enzyme. When the bacteria is sprayed, glyphosate binds to its EPSPS region in a noninhibitory fashion. By inserting this enzyme into the glyphosate-sensitive regions of soy and corn DNA, genetic engineers successfully produced the first glyphosate resistant crops.

The benefits of glyphosate seemed most promising when these crops came on the market in the mid 1990s. Prior to glyphosate resistant (GR) crops, conventional farmers had to use multiple chemicals to eliminate the weeds growing in their fields. While many farmers already used glyphosate to clear fields before planting, they couldn't use it to kill weeds later, because it would kill their crops as well. They had to use other chemicals to target weedy pests instead.

Genetically modified crops containing the new CP4 GR enzyme could withstand applications of herbicide *after* they had emerged from the ground. With this development, farmers could use glyphosate to eliminate all weedy competitors. In addition to making management decisions easier for the farmer, glyphosate-based herbicides also eliminated the need to till GR crops to disrupt weed cycles. Farmers could simply spray crops as needed. In combination, these benefits helped farmers cover more acres while doing less time-consuming work. As yields increased and the government incentivized GR crop production, these crops became cheaper for the consumer and the feed man.

As a result, glyphosate quickly became a popular choice for weed control. Forced to participate within this super competitive system, it wasn't practical for many farmers to ignore GM crops and their chemical counterparts. Less than two decades after Monsanto's *Roundup Ready* corn and bean products were developed, more than 100 million acres of American land were planted to *Roundup Ready* crops. Corn acreage alone saw more than a ten-fold increase in average herbicide use.[4] The paper promises of high yields blinded many Americans to the economic, environmental, and health costs that would come.

The following graph illustrates the rapid adoption of GM crops in the United States. Between the years 1997 and 2012, herbicide tolerant (HT) and insect tolerant (BT) corn, cotton, and soybean varieties went from occupying little to no space to occupying the majority of farmed acres. Most HT crops developed during this time period were modified to tolerate glyphosate applications.

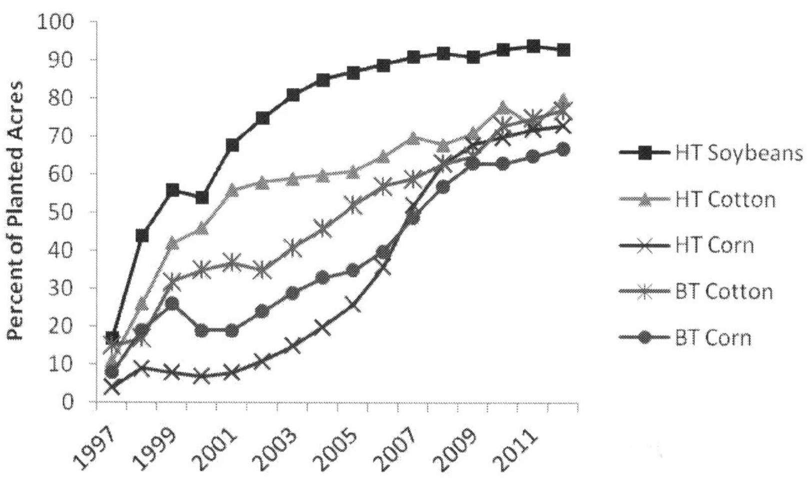

Source: USDA, ERS.[5] Note: Data for each category includes varieties for both HT (herbicide tolerant) and Bt (insect resistant) stacked Traits

During this same time period, national glyphosate use increased five-fold to about 185 million pounds of active ingredient (a.i.).[6] These figures do not include the millions of pounds of glyphosate used annually on private and governmental properties for weed control in lawns, parks, golf courses, gardens, and ditches. Genetic modifications that were supposed to help us rely less on pesticides actually did the opposite.

Estimating the exact amount of current glyphosate use can be tricky due to the sheer number of glyphosate containing products on the market. Since *Roundup* first came on the market, the United States EPA has approved hundreds of chemical products containing the active ingredient glyphosate.[7] These products are sold by a handful of companies and go by a wide variety of names. In addition to Monsanto's original *Roundup* products, you can also buy glyphosate under labels like *Buccaneer, Razor Pro, Genesis Extra II, Roundup Pro Concentrate*, and *Aquamaster*. Glyphosate is often sold as part of a "premixed" formulation—a mixture that contains more than one active ingredient.

As America's genetically modified crop and weed management practices are taken up around the world, we lose our abilities to regulate the unforeseen health effects associated with these pesticide dependent practices. Since Monsanto's *Roundup* patent expired in 2000, numerous equally toxic brands have entered international markets. China, in particular, has become a major glyphosate producer. During the last five years, its chemical factories increased glyphosate production substantially. As a result, global production capacity has, at times, reached twice the global demand![9] Much of this product is used in places like Brazil, China, and India where agbiotech companies encourage politicians and farmers to adopt GM seed technologies. This chemical has truly changed the world of crop farming as we know it!

While promoting herbicide dependent cropping systems, chemical manufacturers have defended glyphosate's use, arguing that it is an environmentally safe alternative to more harmful herbicides. In fact, numerous glyphosate proponents argue that glyphosate, when used in combination with GR cropping systems, adds net environmental benefits to farmed areas.

In a 2010 review called "The Agronomic Benefits of Glyphosate Use in Europe," Monsanto International outlined why glyphosate is the only remaining *viable solution* for *effective weed management*. According to Monsanto's authors, the practice of "conservation agriculture," the technique of integrating glyphosate use, GR crops, and cover crops, increases wildlife and biodiversity, lowers fuel use, and reduces soil erosion. In addition to these benefits, Monsanto has argued that glyphosate is a *benign* chemical that exercises no herbicidal action through the soil or long-term activity on plants that could create public health and safety concerns.[10]

While it is true that glyphosate is *less* harmful, on a per dose basis, than many of the herbicides used previously, all herbicides act as environmental toxins. Herbicides, like other pesticides, are designed to kill target species. They can also inadvertently kill non-target species. Claims that glyphosate can increase biodiversity rely on relative interpretations of science in which it is generally understood that one chemical is better than another, but comparisons to alternative conservation systems are left out altogether. Quite clearly, the discussed "conservation agriculture" techniques rely less on fuel for tilling weeds. It is less clear, however, that these techniques rely more on fuel for manufacturing, distributing, and dispersing chemicals and fertilizers than on systems that make do without additional chemicals.

By the same token, biodiversity is undermined to a much greater extent in these "conservation agriculture" systems. When glyphosate resistant crops are used in modern monoculture systems, soils are repeatedly planted to the same crop and treated with the same chemical year after year. The effects of widespread monocultures on biodiversity are not only obviously visible, researchers have documented and described the less visible, long-term consequences. In many states, such monocultures have completely replaced more diverse ag systems and systems that relied on fewer chemicals overall.

Today crop farmers and scientists are observing a new flaw in "conservation agriculture" models. Of growing concern, is the development of numerous weeds that have evolved resistance to glyphosate treatments. The once touted environmental benefits of this *less* toxic chemical don't seem as promising as the amount of glyphosate needed increases. Each year, the growing number of evolved glyphosate resistant weeds has resulted in increased numbers of glyphosate applications, as well as a need for additional herbicides, putting farmers right back to where they were before this "miracle" herbicide came on the market.

While examining glyphosate's effects on animal health, my father and I have run into an extraordinary number of contradictory claims. Though much of the available literature on glyphosate asserts that it is relatively harmless and beneficial, the bulk of these studies were funded through educational partnerships and grants tied to investor stakeholders.

Recently, several papers have been brought to the media's attention contradicting this literature. This research includes publications indicating that glyphosate negatively impacts soil, plant, and animal nutrition, as well as the nutrition of GM plants and animals eating those plants, in ways that we only partially understand.

Because farmers have generally embraced weed management technologies, studies designed to investigate problems associated with GM technologies and pesticides are seldom funded. While my father has met many scientists who immediately dismiss potential links between *Roundup* and animal health problems, he has also been introduced to scientists who are so concerned about glyphosate-based herbicides that they have devoted their life's work to ridding the planet of glyphosate's use. These scientists are ignoring much easier opportunities to obtain funding for their research. Their work is often marginalized in a world increasingly reliant on agbiotech funds.

So, while it may be true that glyphosate has been reviewed many times, glyphosate's effects on mineral nutrition and animal diseases like manganese deficiency, remain largely under-investigated. This, despite the fact that glyphosate is used so widely throughout the world. These effects may impact various aspects of environmental health, including the health of our nation's soils, the infrastructure supporting the roots of every interconnected, land-based food web.

Theo Colborn discusses the importance of protecting soil health at length in her book, Our Stolen Future.

> Thirty years from now, our children may be struggling to stem another serious assault on the systems that support life. Perhaps the next surprise will show up in the soil, one of the least appreciated parts of our life-support system. The consequences would be dire indeed if human activities were to seriously undermine the soil's ability to recycle nutrients – a process of recycling and renewal that depends on a myriad of bacteria, fungi, and insects (pg 242).[11]

These prophetic words, written just fifteen years ago, follow a presentation of Colborn's disturbing findings regarding endocrine disrupters. Endocrine disruptors are chemicals that interfere with hormone regulation in plants, animals, and humans. After demonstrating how pesticides can negatively affect one's reproductive health, Colburn shrewdly predicted today's most pressing agronomic concern: the declining health of our nation's soils due to widespread chemical use.

Understanding glyphosate's unique ability to chelate a wide range of mineral and nutrient ions helps one understand why glyphosate contributes to the declining health of our nation's soils. Chelating chemicals, like glyphosate, bind with other chemical ions to form inactivated complexes. These inactivated complexes can no longer become important components of various soil-bound life forms.

Glyphosate's ability to bind to soil minerals and nutrients, affecting a *myriad of bacteria, fungi, and insects*, is exactly what makes glyphosate, and other broad-spectrum chemicals like it, such excellent herbicides. When glyphosate disrupts soil-bound organisms' abilities to use chemicals, cellular processes go awry. These molecules are needed to make important cellular components like proteins. With the loss of these components, we deplete overall soil health, especially the soil's ability to recycle nutrients and resist disease.

These alterations later affect crops directly and indirectly. Most plants ultimately die because they are unable to defend themselves from the soilborne pathogens that attack their weakened immune systems. In crops engineered to resist glyphosate, like *Roundup Ready* crops, the modification technology protects the plant to some extent. The plant is not as sensitive to glyphosate so you can apply it directly to the modified plant without killing it. Yet, the glyphosate still acts as a chelator to reduce the uptake of nutrients from the soil.

These chelating actions can decrease chemical and biological processes that convert soil molecules into more bioavailable forms. There are then fewer available nutrients for crops to take up—creating the equivalent of a nutrient desert. When animals and humans later consume these nutritionally depleted plants, they too, may be affected by the accumulating damage done to the soil.

The way this works is complicated. Chemicals like glyphosate have different effects on different soil organisms, and as a result, don't simply render the soil sterile. Instead, they change the mixture of bacteria, mold, and fungi in the soil.[12-14]

Bacteria that do not use the shikimic pathway are less affected by glyphosate than those that do. In a field treated with the herbicide, they will outcompete other bacteria. Similarly, some species of mold tend to respond better to glyphosate than others. When the natural ratio becomes imbalanced, we begin to see negative side effects on the plants grown in this soil. Researchers have reported that photosynthesis, protein metabolism, and root nodulation don't work as well in soils and plants treated with glyphosate. Plants grown in these soils have also been found to be less water efficient and less resistant to disease.[15-24]

While the jury is still out regarding the extent to which such processes affect genetically modified crops, glyphosate's effects on the nutrient efficiencies of non-modified crops is, in many cases, well documented and understood.[25] For example, we now know that glyphosate's activity depends on the plant's weakened immune response. It is actually difficult to kill a glyphosate treated plant in sterile, inactive soil. This is because there are no pathogens to act on the weakened plant.[26] Unfortunately, it is also impossible to grow nutrient rich food in sterile soil. What may be harmful to us, is part of a larger whole. Beneficial and detrimental organisms keep each other in check but it is usually better to have a balance of these species than none at all. One of the reasons we need to supplement animal and human diets with nutrients and fertilizers in the first place is to account for modern inputs that immobilize available nutrients and minerals. Sometimes these inputs wipe out entire microbial species altogether.

Making chemicals in the soil inactive was exactly Stauffer Chemical Company's intention when it marketed glyphosate for this purpose decades ago. In 1964, glyphosate's parent acid was patented as a metal chelator to be used as an industrial waste clean-up.[12] In non-toxic situations, rendering soil particles inactive is seldom desirable.

The effects of soil sterility on human nutrition are explored extensively in numerous publications including the book Tomatoland. In this book author Barry Estabrook attempts to figure out why a commercially grown tomato has 30% less vitamin C, 30% less thiamin, 19% less niacin, and 60% less calcium than it did during the pre-industrial era (pg x).[27] The commercially grown tomato, like many of our commercially grown crops, has been bred to withstand commercially farmed soils and an onslaught of chemicals and fertilizers designed to boost yield and profit. The nutrition of the purchased food product seems relatively unimportant in this context.

Glyphosate presents unique obstacles to soil health because, unlike other herbicides, it binds to a wide range of chemical ions. Without these particles, target organisms wither and die. Unlike target-specific herbicides, glyphosate can bind to both micro and macronutrients in the soil, including calcium, magnesium, copper, cobalt, iron, manganese, nickel, and zinc. These are nutrients that many animals rely on for important metabolic processes.

According to Dr. Huber, this ability also makes glyphosate a "potent antimicrobial agent."[26] Glyphosate doesn't simply affect the EPSPS enzyme, it also ties up any free positive ion, wiping out the bacteria, plants, and fungi that depend on these molecules. Plants grown in different soil types, regions, and with different genetic backgrounds may perform very differently from plants tested in more controlled research settings.

The few researchers investigating glyphosate formulations' long-term consequences are concerned that the influences on subsequent plantings, regional water supplies, and mammals may be much more detrimental than originally believed. Inevitably, these researchers face a barrage of counter-claims from "conservation agriculture" proponents. This is despite the fact that few long-term health and environmental studies have been conducted on both glyphosate-based herbicides and their major breakdown products.[28]

When discussing manganese deficiencies with my father, we've considered that there could be other chemical inputs or nutrients involved that we've failed to recognize. In my father's case, he admits that he has more questions than answers, but gets very frustrated when people are biased against asking important questions in the first place.

To date, he's convinced that glyphosate plays a role in manganese deficiency because he's had success reducing deficiencies when he's removed glyphosate treated feeds from manganese deficient herds.

Unfortunately, as you will see throughout this book, it is very difficult to sort each variable out. Nonetheless, we've become less convinced that many widely used crop chemicals are adding a net benefit to regional food systems. We're also incredibly concerned about the ways such additives may affect long-term animal and human health.

DIAGNOSIS 3: POST WEANING FAILURE TO THRIVE SYNDROME

If you thought you were mistaken when you read the title of this chapter, you are not. One of the many ridiculous names for this swine industry disease is post weaning failure to thrive syndrome. I laughed when I first heard about this disease; it sounds like the complicated scientific name for one's terrible twos or miserable adolescence. Or perhaps they meant to say middle-age failure to thrive syndrome? I've certainly had a case or two of that.

Post weaning failure to thrive syndrome, or PFTS, turns out to be anything but funny. PFTS has also been called post weaning fading pig-anorexia syndrome, post weaning wasting-catabolic syndrome, and post weaning multisystemic wasting syndrome.[1] This disturbing condition has become increasingly common throughout North America in recent decades. Within five to ten days after weaning (the time the piglet is removed from his mother and taken off a milk diet) active seemingly healthy piglets become gaunt, pale, anorexic, and see a rapid deterioration in health. Not only do they "fail to thrive," as the name implies, they often begin to catabolize (break down) their own tissues and organs, essentially self-cannibalizing to survive. In this chapter we explore diagnostic complications related to this condition.

According to a team of veterinarians presenting on PFTS at a veterinary conference in 2008, "Even with early detection and immediate supportive nutrient and antimicrobial intervention, affected pigs catabolize fat and muscle tissue over the ensuing 2-3 weeks and deteriorate to become emaciated and require euthanasia" (qtd in the National Hog Farmer).[2] Veterinarians and farmers can diagnose PFTS but their interventions remain largely ineffective. This seems especially concerning because the syndrome can affect between 2-10% of affected pig operations—sometimes even up to 20% of weaned piglets.[1]

No farmer or veterinarian likes to see 10% of a herd waste away. Sadly, on several operations, such losses are figured into regular "costs." Interestingly, in the numerous studies my father and I have read, figures on the numbers of pigs tossed into the compost heap as a result of PFTS and other common swine industry diseases are largely unaccounted for, inflating industrial claims to have improved overall efficiencies. While I have the utmost respect for producers of quality, well cared for bacon, I certainly don't support practices that might artificially influence any animal's development of mushy bones and deteriorating organs. This does not make me a crazed animal rights activist, as the swine industry might have you believe. Like others concerned about these problems, I've simply managed to retain my sense of right versus wrong. Numerous farmers and veterinarians are equally concerned about PFTS on their farms.

The question then is: have we in some way created this problem or is it beyond our control? After careful observation, we've come to the conclusion that swine industry-funded research institutions are overlooking valuable clinical information due to the context in which they work. Broadening research angles could help us find answers and much desired treatments in a more timely manner. Again, we make a series of deductions based on observable evidence and admit that we have more questions than answers.

Though many originally thought that a virus might cause PFTS, recent research found "a lack of compelling evidence linking [PFTS] to common pathogens."[3] According to veterinarians involved in this research, PFTS may be the result of an infectious agent such as porcine circovirus or other factors including nutrition and management. Pigs with Porcine Circoviral Disease (PCVD) often exhibit symptoms of PFTS but the relationship is unclear. Likewise, PFTS does not consistently occur after a pig is infected with common hog viruses, such as PRRS, PCVD, or rotavirus, but does appear to be worse in pigs with these conditions. These observations do not eliminate the possibility that a virus plays a role in PFTS, but they do lead scientists to consider other factors.

Because disease onset occurs when a piglet is weaned and starts eating feed, it seems natural to question dietary roles. To lend credit to this theory, veterinarians often find lesions in the pig's stomach and intestines when a pig is posted.[2] Interestingly many PFTS pigs stop eating supplied feed sources altogether. Their tissues then waste or catabolize as they begin to digest their own tissues for food.

My dad has posted several PFTS pigs with digestive system lesions. These pigs have altered intestinal linings with shrunken villi and thickened crypts—conditions that lead to absorption problems. Farmers also report that their pigs are pale and anorexic before they die. Pigs with PFTS are either born with different intestines or quickly develop different intestines as a result of the syndrome.[4]

This information, combined with the knowledge that piglet feed quality has decreased in the last decade, further supports the view that PFTS may have an unknown dietary cause. According to an article in the *National Hog Farmer*, several nutritional factors could play a role. A swine veterinarian interviewed by NHS, Steve Henry, forwarded that cost-savings moves made by the industry could impact PFTS. He identified several areas in pig starter diets that have seen recent changes. These include: lower quality whey in starter diets, a decline in the quality of fish meal fed to weaned piglets, reduced feeding budgets for starter diets, and an increased number of ingredients used to improve performance and reduce costs.[4]

When I asked my father about this list, he agreed with Henry's concerns and expanded on several points. In the case of reduced feed quality, he explained that increased demand for nutrient supplements in human markets has decreased availability of quality, affordable nutrients for many animals. Nutrient supplements are another input industry-funded studies often overlook.

He also mentioned that imported, synthetic vitamins may be more available and affordable in the market but less bioavailable to the body. Today, the animal feed industry, when vying for quality feed supplements such as fish oils and natural vitamin E, has to compete with the human food industry. Animal nutrition is sometimes compromised as a result. This competition has increased in recent years as American nutrient suppliers realize that foreign sources of nutritive supplements are not always up to par with consumer demands.

Just as human doctors are concerned about the increasing number of innovative food-stuffs in packaged food products, veterinarians need to take all chemical and feed inputs into account when treating diseased pigs. Unfortunately, the number of variables in pig starter diets presents challenges to any diagnostician. To achieve maximum weight gain, ingredients are often added to traditional feeds. Not only do they sometimes compromise overall animal health, these additives increase the number of factors tied to herd health problems. Amongst the several nutrient deficiencies that could play a role in PFTS, Vitamins E, D, and Iron are all potential candidates.

To complicate matters, it has become increasingly common to feed pigs corn distillers, an edible waste product of the ethanol-manufacturing process, a practice that is most practical when corn prices are low. Feeding animals corn distillers benefits reported ag-business efficiencies in two ways: first, by reducing overall feed costs to the farmer and second, by increasing the profitability (and therefore apparent efficiency) of the corn-based ethanol-making process. Unfortunately, the nutritive quality of corn distillers is not as high as traditional feed sources. This was evident in one serious PFTS case within my father's practice where farm operators were feeding a very large amount of corn distillers to weanlings. These diets contained 600 pounds of corn distiller per ton and were supplemented with synthetic amino acids. Prior to the availability of these alternative feedstuffs, piglets were fed a more traditional diet including high-protein bean meal.

When considering the implications of a corn distiller diet, a thorough scientist must consider a range of new and complicated variables. Like our industrial tomatoes, the seed corn that becomes corn distillers, does not always harbor the nutritional value it used to. According to a comparison of 14 Iowa research sites, a bushel of corn in 2006-2007 contained .56 pounds of nitrogen per pound compared to .82 pounds in the 1970s.[5] This is more than a 30% decrease in the amount of nitrogen per bushel of corn—a troubling figure given nitrogen's role as an important building block in body proteins.

In addition to the questionable nutrient composition of a distiller-based diet, distillers are typically made from GM corn varieties that have been repeatedly treated and sprayed with a variety of pesticide products including glyphosate. If we ignore all other pesticides (of which there are many) and simply focus our attention on glyphosate you get an idea of how complex diagnosing disease has become. Keep in mind that most industrial diagnosticians don't even consider genetically engineered crops and pesticides potential risk factors for disease because our FDA has deemed them safe for consumption.

Even with our limited experience and resources, we have identified several potential links between PFTS and glyphosate-based herbicides that seem worth considering. In fact, my father has become so frustrated with the lack of research in one area that he is trying to fund his own rather extensive research project. My mother, bless her heart, tolerates this while remaining somewhat anxious about their ability to retire and afford decent health insurance.

Let's start with the most familiar concept: pesticide-induced nutrient deficiencies. As we read in the last chapter, glyphosate can bind to micro and macro nutrient particles. Its action is not limited to manganese alone. In fact, researchers at Texas A & M found that glyphosate ties up cobalt at 10^2-10^3 times more than it ties up manganese. The same liver lab analysis my father used to confirm manganese deficiencies also revealed rock-bottom cobalt levels in tested hogs.[6] Out of 522 livers tested, none hit the normal range for cobalt established in the pre-GM era.

Unfortunately, testing cobalt levels was previously so uneconomical that cobalt deficiency has not been on veterinarians' radar. My father only became aware of these cobalt deficiencies after Iowa State University purchased a plasma mass spectograph. Prior to this purchase, it cost him $27 for each trace mineral test he ordered. Today, he can test numerous trace minerals for the relatively low price of $56.

Cobalt could play an important role in PFTS because it is essential to forming vitamin B_{12}. Some veterinarians suspect that low B_{12} could influence PFTS because B_{12} helps many animals digest carbohydrates. When pigs don't get enough cobalt, they don't make enough B_{12}. As a result, they have a difficult time digesting carbs.

Humans and pigs are physiologically very similar in this regard. The heavyset kid who is told to exercise and eat better will still be unable to handle carbohydrates if he is B_{12} deficient. This puts him at risk for Type 2 diabetes and a range of other illnesses later in life. Researchers studying rising incidences of diabetes in American children might be interested to learn that diabetes is on the rise in cats, dogs, and pigs as well. While researchers have demonstrated that human's corn-based sugar dependencies are likely to blame, B_{12} deficiencies could also play a role in the increasing numbers of Type 2 diabetes.

Unlike the humans contracting Type 2 diabetes, the pigs exhibiting PFTS aren't eating refined sugar. They're going from a milk protein based diet to a carbohydrate based diet and seem either unable or uninterested in digesting this grain-based feed. When dosed with oral B_{12} supplements, my father has observed some improvements in wasting piglets, as well as in sow reproductive measures, especially amongst pigs bedded on glyphosate treated cornstalks.

As I mentioned earlier, animals need cobalt to form vitamin B_{12}. Animal protein content is typically our largest source of cobalt, but pigs eating grain-based diets have to rely on grain sources for cobalt intake. When cobalt is unavailable, plants can suffer from cobalt deficiencies too.

In fact, cobalt's role in protecting nitrogen-fixing plants, like soybeans, can be compromised following glyphosate application. The glyphosate somehow interferes with normal cobalt uptake.[7] Research conducted in 2009 demonstrated that cobalt is not only needed by the bacteria in soybean root nodules, it is also needed by all plants to protect them from root rot.[6] A lack of cobalt makes a plant more susceptible to this disease.

In pigs, B_{12} absorption is also influenced by intrinsic factor. Intrinsic factor is a protein secreted by the stomach that's necessary to absorb B_{12}. If your body is unable to produce intrinsic factor, it is also unable to absorb vitamin B_{12} even if you are given oral B_{12} supplements. What if something about the GM diet or glyphosate's antibiotic action irritates the stomach lining reducing one's ability to produce intrinsic factor?

Interestingly, one of the studies comparing rats fed GM diets with rats fed non-GM diets lends credit to this theory. Many of the rats fed GM diets exhibited stomach and intestinal lesions similar to pigs with PFTS.[8]

While considering these factors, my father has noted differences between the ways pigs and ruminants (like cows and goats) digest their feed. He thinks that glyphosate treated, genetically modified feedstuffs may affect piglets more than ruminants eating similar feed items because they digest these foods differently. Grazing ruminants may be less susceptible to B_{12} deficiencies because their four chambered stomachs are packed with bugs that help them break down toxins. Non-ruminants like pigs and humans don't have as many bugs in their guts to do this work. Intrinsic factor may not play as important a role in absorbing B_{12} to these ruminants. Certainly, more information is needed to ascertain what effect pesticides have on the important bacteria in our guts. Most of the studies conducted so far are limited to rats and rabbits.

In addition to these nutrient absorption concerns, management and environmental practices may play a more significant role in PFTS than is widely recognized. While most reviews of this disease seem to acknowledge that management and environmental factors may influence PFTS, few studies actually evaluate these basic factors. Researchers seem to assume that pigs can only be raised in densely packed confinement units.

In their peer-reviewed paper, "Clinical presentation, case definition and diagnostic guidelines," Huang et al. acknowledge that research is lacking. They find that "most management interventions, environmental manipulations and medical treatments are ineffective," and that "a critical review of all production practices must be performed to reduce stress, to prevent exacerbation of all nursery diseases including PFTS, and to ensure animal welfare is maintained. Techniques that are consistently effective in reducing PFTS morbidity and mortality have not been accomplished."[1]

Such conclusions seem interesting in light of the comments made by a veterinary team in the *National Hog Farmer*. According to this team, "it can be difficult to identify and solve this syndrome in a normal research setting because it's tough to replicate what's happening in the field. Providing extra care in the research barn to carry out the trial often results in improved performance."[2] If this is so, wouldn't one naturally suspect that environment and management have something to do with the PFTS problem? What exactly do researchers mean by *most management interventions, environmental manipulations, and medical treatments are ineffective*? Which interventions are excluded from experiments, if any?

According to this same article, environmental practices such as air conditioning to optimal performance levels and post-weaning competition may put pigs at greater risk. This article also reports that supplementing feed with electrolytes, nutrients, and milk products can be "beneficial, but more often confusing."[2] Apparently, some pigs begin to feed again but don't fully recover.

While researchers are sacrificing hogs to evaluate internal organ damage and spending big dollars on blood tests, I could find no peer reviewed information comparing on-farm management practices. I did find an online fact sheet published in Great Britain providing information for farmers and veterinarians in the event PFTS pops up in the UK. According to this information sheet, some specialists "have expressed the view that PFTS relates mainly to suboptimal management of pigs in the immediate post-weaning period, resulting in some pigs not finding food and 'starving out.'" It also seems important to note that one difference between the U.S. and Europe is that "pigs can be weaned earlier than 28 days old in the US, while this is illegal in the UK unless special provision is made."[9] The authors of the fact sheet listed infections, viral diseases, overcrowding, chilling, and feed issues as possible causes of PFTS. The feed issues described included deficiencies and mycotoxins (toxic mold residues that develop on feeds).[9]

While I realize that it may seem insulting to question management and environmental practices, I think most conscientious farmers would want to know if their management practices are affecting PFTS and the losses involved. Even in my limited time investigating this problem, I've found a need for the following animal research studies:

1) Studies comparing the prevalence of PFTS in pastured hogs and confined hogs to confirm that cooling and population density are not primary agents in PFTS.

2) Studies comparing pigs fed GM feed vs non GM feed to confirm that something in GM bean and corn meal is not affecting the gastro-intestinal lining, including the pig's ability to produce intrinsic factor.

3) Studies comparing non GM, GM, and distiller diets to

eliminate dietary quality issues including the potential role of mycotoxins in making a pig more susceptible to PFTS.

4) Studies comparing piglets receiving vitamin B_{12} injections to those who are not.

5) Studies comparing blood content levels of key nutrient suspects including cobalt, vitamin E, and vitamin D in both newborn piglets and recently weaned pigs.

6) Studies demonstrating how weaning age may or may not affect one's ability to thrive in a confinement unit.

In the case of PFTS, only one thing is certain. Despite the fact that we are learning so much about increasing consumption and economic *efficiencies*, getting pigs started on grain diets remains a problem. Are our current research efforts for naught?

Given the similarities between pigs and humans, it seems especially important that we look into diseases like PFTS. Pig health problems are often indicative of human health problems. Pigs are so biologically similar to humans that we've been able to use pig valves, pig skin, and pig organs in human medicine and research. Humans, like pigs, rely on absorbed nutrients from the stomach and intestine. We are what we eat in a very literal sense.

According to doctor Mark Hymen, author of The UltraMind Solution, nutrient deficiencies are at the root of many American illnesses and have contributed to a society that increasingly relies on antidepressants, stimulants, and memory medicines to deal with ADHD, depression, mood disorders, autism, and alzheimers. Hymen explores how such disorders are linked to our physical wellbeing, as well as epidemics like diabetes and obesity. He claims that a mere 10% of Americans are nutritionally, metabolically, and biochemically balanced enough to benefit from mental therapy (pg 13). He also explains how serious vitamin B_{12} and omega 3 fatty acid deficiencies affect a large portion of the American population. Instead of getting ourselves hooked on unnecessary drugs, he proposes that Americans address the root of the problem: our diets.

Though Hymen's findings predominantly focus on human nutrition, his research, in this case, is somewhat similar to my own. The same deficiencies affecting humans can affect cows and pigs. Hymen and other like-minded doctors encourage Americans to identify what they are missing and start appropriate supplement and detoxification regimens to fix their ailing systems. Veterinarians, likewise, evaluate animal nutrition problems and encourage producers to supplement animal feed rations.

What disturbs me the most about such prescriptions is that they still don't address the true roots of such problems: the soil that is no longer healthy enough to sustain nutrient rich plants, the food on which herbivores, carnivores, and omnivores ultimately rely. While scientists have explained how toxins can bioaccumulate in a food chain and contribute to the "pesticide treadmill," our ever-growing need to apply more and more pesticides to address the same pests, we seem largely unaware of another input-intensive component of this system altogether: the nutrient supplement treadmill.

Supplementing all plants, animals, and humans with nutrients that used to occur naturally, does not come without costs. Yet this is the oft proposed solution to our industrial ag ailments. In fact, many of us already rely on this treadmill. We even talk about nutrient supplementation like it is a good thing, when in fact it is an unnecessary cost that can be a bad thing: one more source of unreliable variables mined or bioengineered in God-Knows-Where, Thank-You-Globalism. A nutrient supplement is not only a feedstuff, it is a feedstuff that can contain multiple geno-chemically modified ingredients from various locations around the world.

While I am fully aware that it is sometimes necessary to take nutrient supplements, I do not want to become more reliant on them in the future. We are what we eat in the truest sense—our diets being the largest variables we can control when it comes to human health. By eliminating unnecessary toxins in our food supply, we can retain more of this control.

Investigating PFTS has helped me realize that our food is much more than calories. It contains the basic information we use to direct our genes. Without healthy food, we can waste away. Our organs can eat themselves and our bones can turn to mush. Yet, year after year, we apply chemicals and fertilizers to our soils that tie up valuable bits of this information and we lose a little bit more of our national health. While those who can afford to purchase quality dietary supplements may preserve their qualities of life, too many others will suffer from ambiguous symptoms linked to a nutrient deficient diet.

The ability to afford quality supplements will also become increasingly difficult as demand grows. Veterinarians are already seeing this with supplement prices in the animal sector. Ironically, we find ourselves competing with animals for quality supplements even though we later rely on these same creatures for food.

We cannot continue to ignore such costs. The animals we eat should not be dying of nutrient deficiencies within the contexts of their already shortened lifespans. They are doing us a service and deserve better. Likewise, our children shouldn't need to belong to the upper class to enjoy nutrient-rich foods. Expecting raw foods to be nutrient rich and more affordable than the processed, highly modified foods and animal feeds we use today is a worthwhile goal. Looking into the true causes of diseases like PFTS will help us achieve this goal.

DIAGNOSIS 4: INFERTILITY

When I started researching relationships between plant, animal, and human diseases, I had no idea I might one day write this book. I was six months pregnant, teaching full time, and over-committing myself to projects as usual. It wasn't until after a scare with my newborn that I more fully committed myself to this project. After six weeks of thinking I'd birthed a perfectly healthy baby boy, my husband and I were terrified to discover that his liver was malfunctioning and that he might need a transplant.

We spent the next two weeks panicking, seeing specialists, praying to various Gods, and driving our ten pound bundle of joy between laboratories, hospitals, and imaging units. After obtaining a mere one hour of sleep in two weeks, I received joyous news: our son didn't actually need a liver transplant; his bile ducts were simply engorged with tiny stones – a product of an environmental toxin or developmental complication. He could have a surgery and, if all went well, go home.

After too many hours of anxiety-filled decision making, I couldn't believe our luck: surgeons were able to dislodge the stones using a modified scoping device in a closed operation. In a small room in Boston Children's Hospital, I collapsed on the floor and wept. It felt so good and so bad at the same time. Good because I was so relieved! Bad because we were surrounded by other parents who hadn't received such good news. Parents whose entire lives were altered by their children's high-maintenance ailments. Previously numb, I felt like crying for them too.

And that's about the time I started to feel really mad. Prior to my son's troubles, I had been helping my father sort through research pertaining to various illnesses linked to animal feed supplies. But I had never really stopped to imagine the true implications this research might have for our nation's children and parents. "How much empirical evidence do we need to deem a chemical unsafe?" I wondered. "How many links to health problems must we establish before a chemical is banned from use or limited in the marketplace?"

The straightforward answer is way too much. Infants, in particular, are exposed to far too many dangerous chemicals before they are born. One study found an average of 200 industrial chemicals in infant umbilical cord blood. Of the 287 chemicals found, 180 are linked to cancer, 217 are toxic to the brain and nervous system, and 208 cause birth defects or abnormal development in animal tests. Furthermore, many of the dangers associated with these toxic mixtures haven't been studied.[1]

This news is particularly alarming to a new parent. Apparently, researchers only have the capacity to test a small percentage of the tens of thousands of chemicals we currently produce.[2] Even when federal agencies are trying their hardest to safely regulate toxins, they must operate within budgetary constraints. They often can't keep up with the overwhelming number of new chemical combinations on the market.

At the rate current political change takes place, our soils and wombs will be completely sterile before we acknowledge a need for change. How many of the unborn can live healthier, less worrisome lives if we take better care of the land around us? How much land will needlessly become an ecological and agricultural dead zone, fueling future needs to modify the landscape in ways that unknowingly promote harm?

My father often wonders these same questions. During the last ten years, he's noticed that cattle and pigs eating *non*-GMO, *non*-glyphosate treated grains during the breeding season have significantly higher fertility rates than cattle and pigs eating pesticide treated, genetically modified corn and bean meal. In some cases between 15 and 25 percent of a herd have an early term abortion or fail to breed. Just twenty years ago, these numbers were between 1-10% throughout his practice.

Infertility and fetal abortions are, in fact, well discussed problems within the veterinary world. In a November 2011 issue of *Hoard's Dairyman* I read an article entitled: "Why are so many cows losing pregnancies? Losing up to 20% of pregnancies is not acceptable." According to the authors, experienced cattle breeders are reporting increased numbers of fetal abortions despite increases in semen costs, nutrition costs, and overall breeding costs throughout North America.[3]

The problem is so bad that the U.S. Cattlemen's Association made a statement to Congress reporting that, "Cattle ranchers are facing some puzzling - and, at times, economically devastating problems with pregnant cows and calves. At some facilities, high numbers of fetuses are aborting for no apparent reason. Other farmers successfully raise what look to be normal young cattle, only to learn when the animals are butchered that their carcasses appear old and, therefore, less valuable."[4] In a few herds, up to 40-50 percent of pregnancies are lost. It appears that young pigs are not the only creatures failing to thrive. In addition to other challenges, livestock farmers often need to address herd fertility issues as well.

My father seems convinced that many fertility issues can be fixed by improving the animals' feed supplies, specifically by reducing the amount of glyphosate treated GM corn and beans used to feed animals. Unfortunately, proving that reproductive issues are influenced by glyphosate treated feeds is super complex because there are so many factors that typically influence an animal's fertility. These include various aspects of an animal's genetics, environment, and diet.

Limitations inherent in selecting an animal's genetics certainly play a role in an animal's ability to successfully reproduce and digest its feed. As breeders attempt to genetically enhance a cow's productivity, the genetic costs associated with increased milk production and weight gain have begun to outweigh other benefits. According to Stan Bevers, an extension economist, average beef calving rates have declined 1.3%, calf weaning weights have declined 36 lbs, and the pounds weaned per cow have declined 25% in the last few years.[5] This is despite industry's attempt to improve other aspects of the cow's genome via selective breeding.

Natural selection is optimized in a diverse environment. Those cows with the strongest combination of genetic traits survive to pass their traits on to the next generation. However, in our current system, where a few traits are prioritized above all else and the animal isn't expected to live beyond a couple of years, natural selection doesn't have the time or gene pool necessary to do this important work. The cows are no longer genetically and chemically matched to their feed supplies, which today include a number of genetically modified plants.

To be fair, part of the problem is nothing new. Humans have been selectively breeding animals and plants for centuries: picking out the most valuable traits, breeding these traits, and reaping the benefits. By selecting for traits that benefit humans in certain ways, we have often ignored the importance of maintaining a diverse and robust gene pool.

However, several of the problems we associate with selective breeding have been greatly accelerated by modern biotechnology practices. New technologies have allowed us to more quickly influence certain aspects of genetic lines while giving even less regard to others. The co-evolution of an animal and plant that used to take place over thousands of years is replaced with laboratory processes that take minutes. Many of these processes would never take place in nature, even had the organisms been granted tens of thousands of years to mate in their natural habitats.

Some selective-breeding biotechnology practices are less invasive and risky than others. In addition to traditional selective breeding programs, biotechnologists today can use marker assisted selection (MAS). MAS is a relatively new practice in which geneticists speed up selective breeding by identifying genes linked to important traits. Animals with these traits are then bred. Because no one is physically manipulating an organism's DNA, MAS is considered a fairly non-controversial type of biotechnology.[6]

Biotechnology practices that receive more attention in today's media include genetic modification technologies and cloning. A genetically modified organism (GMO) can be created when scientists use recombinant DNA (rDNA) techniques to transfer genetic materials between organisms. This transfer would not occur naturally. Because this technology alters the genetic make-up and expression of an organism, the use of GM technologies is not as widely accepted. GM plants, like Roundup Ready corn and beans, are increasingly common in the American feed supply and Western diet. GM animals, like genetically engineered salmon, are also being developed for introduction to human-food markets.

Most of the GM organisms developed in the last twenty years were developed to tolerate herbicides, as we discussed in Diagnosis Two. Nonetheless, GM organisms have also been engineered to express a variety of other traits. These include increased transportability, insect resistance, and altered fertility. As you can see in the following chart, glyphosate resistant crops are just one type of GM organism common throughout human and animal feed supplies.

GM Food Examples		
Desired Trait	**Sample GMO**	**Intended Gene Function**
Herbicide Tolerant (HT)	RR/HT soybeans	Withstands herbicide applications that weaken & kill weedy competitors
Insect Resistant	BT corn	Expresses bacterial pesticide gene *bacillus thuringeineis* (BT) to kill insects feeding off crops
Virus Resistant	Papaya	Contains viral-coated proteins resistant to papaya ringspot virus
Delayed Ripening	Tomato	Supresses enzyme to delay ripening for transport
More efficient nutrient use	pig (in R&D only)	Produces enzyme phytase to release indigestible plant phosphorus via saliva

*Note: This list is far from comprehensive. There are many types of GM foods in our food supply and many animal modifications under current investigation. For a more complete & updated list of GM foods under review & in our food supply see the USDA's Chart on Petitions for Non-regulated Status.

While we have certainly improved our abilities to select for specific traits, both via increasing the speed of traditional breeding techniques and via rDNA technologies, the short-term successes of such projects are creating long-term consequences. With regard to fertility and abortion problems, my father suspects that we are outdoing ourselves: modifying both the environment and feed intake to such extents that a cow's stomach can no longer properly digest available food, what's more, carry a calf full term!

A common assumption, however, is that these fertility issues can be blamed on genetics alone. This is because the problems associated with genetic inbreeding are widely recognized. In this chapter we will explore how increased rates of both gene selection and gene manipulation affect fertility measures. To begin, its important that you have some background knowledge of plant and animal breeding methods.

Today most of my father's clients use an artificial inseminator to breed their beef and dairy herds as opposed to an on-farm bull. Artificial Insemination (AI) facilitates rapid genetic selection by industry-approved, top quality bulls. AI is used so widely in the United States that some highly ranked bulls have sired more than 100,000 offspring!

This practice is becoming more popular around the globe. Oddly, I now have a friend who is an international semen saleswoman, a position that simply didn't exist fifty years ago. Artificial inseminators and semen salespeople can help animal farmers select desired traits. Amongst the many interesting agribusiness magazines my father receives each year, there are always one or two breeding specials featuring individual shots of highly regarded bulls. These subjects are pictured above an impressive list of productivity stats. Quite honestly, these magazines are not very different from an on-line dating catalogue. If you purchase the semen of your dreams, your herd will surely be the most productive it's been in years!

Professional breeders and farmers may decide to check a bull's fertility if they are experiencing herd fertility issues. In this case, a veterinarian must collect a semen sample. This is no small feat, as I learned at an all too early age while assisting my father on vet calls.

My husband, born and bred in the Boston suburbs, wasn't quite anticipating this aspect of rural education when he agreed to go on a vet call with my father the second time he journeyed to Iowa. My dad, who loves company on the job, took Ben to treat a milk fever before journeying to another farm where one of several jobs on the to-do list was to "fertility test the bull." My father, being the mischievous creature he is, decided it might be pretty amusing to make his daughter's urban boyfriend help collect a bull semen sample—an often lengthy process that requires additional hands. My husband, who only went on the call to get to know my father better, had no idea what was in store for him!

There are three commonly used techniques for collecting animal semen in the United States: manual manipulation, the artificial vagina technique (AV), and electroejaculation. The technique of choice largely depends on environmental hazards, the species being collected, and the disposition of the animal.

Due to environmental hazards and the bull's generally uncooperative disposition, my father eventually decided he would need to use the electroejaculation technique to obtain the bull's semen. In this case, the veterinarian applies a series of short, low-voltage pulses to the pelvic nerves of the bull to elicit the desired response. The tools used consist of an electroejaculator connected to rheostats to control and manipulate the current, a collection cup, and a torpedo-shaped probe.

As you might imagine, Ben, who had only been on a handful of farms up until this point, was not prepared to assist with a bull semen collection process, especially given the rather strange, somewhat dangerous, and unexpected nature of the job. My father, who is accustomed to stubborn animals, talked about the process like it was old hat. He then strung a rather intimidating looking contraption, including attached wiring and cables, around his neck, before inserting a nine-inch probe up the bull's rectum and finessing the rheostat's knobs. My husband later reported that he looked like a science fiction villain. My father did all of this while instructing my future husband to get ready to catch the semen in a collection cup!

All I can say is that I'm so glad I had no idea what Ben was up to until after the job was done! Four years later, my siblings, who teased Ben mercilessly after hearing his rather comical interpretation of this particular vet-call, presented Ben with a bull probe and plaque at our wedding. They did this in honor of the many awkward obstacles he had to overcome to get to know my father! Quite frankly, I'm surprised experiences like this didn't reduce my own fertility success rates!

Because my father has had a chance to monitor fertility successes and failures in herds for several decades, he has been privy to the unforeseen consequences of rapid trait selection. In the same way that inbreeding can increase one's chances to possess rare genes or cause disease, selective breeding and genetic modifications can put animals at greater risk for health problems affecting their bodies and behaviors. While these problems are more obvious in well-monitored animal breeds, researchers are beginning to analyze the costs of rapid trait selection in plant breeds as well.

In certain species, the costs of selecting for a specific trait are difficult to ignore. This is why my father loathes dachshunds. Selective breeding for disproportionately short legs and long backs has led this entire dog breed down a road to chronic back problems most veterinarians find difficult to fix. Every time he's had to treat a dachshund it's for a ruptured vertebral disk. There's not much you can do to prevent this kind of injury when your body looks like a sagging jump rope.

He also gripes about many varieties of purse dogs. Because they have difficulty bearing weight on such fine bones, many miniature purse dogs are more likely to experience slipped and dislocated kneecaps. These conditions are directly linked to the features breeders sought to perpetuate: in this case, reduced body size.

So what happens when a nation tries to increase an animal's milk productivity by 25% in a mere 25 years? When my father started his veterinary practice in the seventies, the average dairy cow in his practice produced roughly 50 pounds of milk per day. Today many farmers expect their cows to produce 85-90 pounds or more. For years, industry boasted efficiencies gained in milk production while largely ignoring other aspects of a cow's health. Increased incidences of lame feet, reduced fertility rates, and predispositions to bouts of milk fever and mastitis are a few of the conditions veterinarians associate with increased productivity demands.

The unintended health consequences of selective breeding beg yet another, more pressing, question: what happens when you insert a gene into an animal's body to resist disease or produce a molecule it wouldn't otherwise produce, as many scientists currently propose we do with genetically modified cows, salmon, pigs and other animals raised for food? If we alter a cow's genome to produce more milk in general, milk with higher vitamin content, or milk with fewer potential allergens, what is the cost to the rest of the animal's body? Where will the proteins these new, improved genes make come from? What other body systems will be affected by these gene insertions?

When a cell is engineered to produce a new protein, as is the case in corn engineered to express insecticidal proteins, the cell requires building blocks to manufacture these proteins. These building blocks must come from within the organism itself or from our soils, demanding additional chemical inputs. When every cell within each of 30,000 GM corn plants per acre demands these extra protein building blocks, there are necessary environmental costs. These costs might include water use (if the proteins consist of additional hydrogen and oxygen atoms), nitrogen use (if the proteins contain additional amino acids), or other required inputs. Whenever we increase the genetic potential of an organism, these changes must draw on chemical resources from within the organism or the ecosystem as a whole.

Even genes that are naturally used cannot act without appropriate chemical resources to create the proteins they blueprint. Past research indicates that genetic insertions in plants may reduce a plant's nutrient uptake and efficiency, as well as its susceptibility to drought and disease.[6-10]Is it also possible then, that these insertions could unintentionally reduce an organism's reproductive potential?

While selective breeding has helped society reap many benefits, my father is convinced that modern ag-biotechnologists, looking for quick fixes to economic problems, aren't always acknowledging the many ways an animal's body must adjust, and over time, adapt, to meet increased productivity demands. These additional demands may be affecting the reproductive health measures of livestock herds in ways that are difficult to assess.

When these demands are made of entire animal populations and farmers use the same bulls to sire thousands of heifers on a handful of large dairies scattered across the world, the incestuous nature of our food system creates additional costs: the reduction of the reproductive gene pool. According to the Food and Agriculture Organization (FAO), about 22% of the world's livestock breeds are currently at risk of extinction. The FAO has urged world governments to put programs into place to reverse the alarming declines in the number of indigenous livestock breeds around the world.[11] These breeds are agriculturally valuable because they are well adapted to local environments and affordable in low-income regions. More importantly, they contain genetic resources for future breeding programs.

When we looked into it, we discovered that the lack of genetic diversity is already affecting the reproductive success and profitability of the United States dairy industry. Researchers at the University of Wisconsin-Madison examined the effects of genomic inbreeding on milk production, reproduction, and udder health in Holsteins, the most prevalent breed of dairy cattle. Results from their work indicate that for every 1% increase in genetic inbreeding, farmers lose 46-116 pounds of milk per lactation and gain 1.06-1.76 days between calving and conception. These figures indicate that it may be harder to breed these cows back after calving.[12]

Still, these costs can't possibly account for all of the observed declines in reproductive health measures. In addition to accelerated breeding technologies, genetically modified feeds may affect reproduction. Like several other farmers and researchers, my father has seen improvements in herd health after animals were removed from GM bedding and switched off GM feed sources. These observations create another important fertility measure question. Are these fertility issues a result of the GM technology itself or the chemicals used so regularly on GM feed and bedding?

While investigating these matters, my father came across a book by Jeffrey Smith, author of Seeds of Deception. Smith has spent several years documenting the damaging effects of GM seed technologies. He currently heads several statewide initiatives to press political leaders to better regulate and label genetically modified foods.

In his publications, Smith outlines several international studies linking fertility problems to the consumption of genetically modified foods. These reports describe separate Asian, European, and African incidents in which animals fed GM diets suffered from higher than normal reproductive issues.[13-18]

These issues included high infant mortality rates, decreased fertility rates, reduced litter size, and reduced pup size amongst mice and rats fed GM diets[14,19] Similarly, a group of Indian veterinarians reported that buffalo eating GM cottonseed had frequent abortions, prolapsed uteruses, and increased incidences of infertility.[18] Since reading this research, my father has visited with several American and Canadian veterinarians who are asking similar questions: what aspect of our chemically coated, genetically modified diet is exacerbating observed fertility problems?

To many veterinarians such reports seem more confusing than anything. This is because our Food and Drug Administration has deemed genetically modified foods and animal feeds *safe* and *substantially equivalent* to non-modified feed crops. They are also interesting in light of the number of variables involved. Unfortunately, because the natural environment and feed supply are already so altered, it is difficult to tell what percentage of these reported statistics are linked to the gene technology itself, as opposed to the pesticides and supplements added to these feeds.

Matters are further complicated by the fact that, historically, researchers in the United States have had limited access to patented seed technologies for side-by-side comparison studies. This access could unveil differences between genetically modified crops and their identical "isoline" varieties. While GMO proponents, including feed-the-world philanthropists and powerful biotech industries, claim that GM crops are safe, nutritious, and necessary, a number of researchers wonder how such claims can be made before researchers are granted greater access to patented seed technologies. Unfortunately, the widely held view that GM crops are "substantially equivalent" has been foundational in the approval and regulation of GM technologies in human and animal foods.

Some scientists tried to warn the public that the principal of substantial equivalence was scientifically flawed before GM seed came on the market. According to Dr. Patrick Brown of the University of California Davis, the presumption that recombinant DNA techniques would not produce harmful effects contradicted literature at that time. This literature documented *"unexpected and often lethal perturbations"* including *"metabolic and phenotypic variations"* in research organisms.[20] Brown argued that, contrary to marketed messages, scientists knew that recombinant DNA technologies were less reliable and precise than the natural breeding techniques humans relied on for centuries. Today we know that these effects may include altered nutrient content, a need for increased inputs, elevated immune responses, and the disruption of entire ecosystems.[6-10]

In order to understand how these variations come about, one must have some understanding of the technologies used. We'll briefly explore the two most widely used transgenes in modern agriculture: those that increase herbicide resistance and those that are modified to contain an insecticide. Then, we'll demonstrate how the variations produced can lead to some of the aforementioned problems.

Whereas herbicide resistant genes are linked to increased demands for herbicides, some insect resistant crops do not rely on external chemicals to achieve desired ends. The insecticide is essentially built into each plant cell. As the altered seed divides, the modified portion of the genome becomes incorporated into the DNA of every plant cell.

When a target insect feeds on this plant, the altered gene binds to receptors in the insect's gut causing cells to deteriorate and die. In this way, genes that confer insect resistance (like the Bt gene) are supposed to reduce insecticide use. They've been designed to kill target insects without the need for additional chemicals.

Unfortunately, this once promising technology has had several undesired consequences. Unlike most chemicals (which eventually degrade in an environment), a manipulated gene cannot be removed. New research demonstrates that these new environmental components can increase insect resistance, affect non-target species, and wind up in non-modified varieties via the distribution of pollen.

Like glyphosate resistance, Bt resistance has been documented in both the lab and the field. To date, at least six populations of insects are resistant to Bt crops. Several of our farmer neighbors cannot believe how much more insecticide they've needed in recent years to kill resistant species. In fact, last year, if you felt jaded paying both a crop protection fee (for your Bt corn) and additional insecticides to kill the bugs the Bt was supposed to kill, you could file paperwork with Monsanto to request compensation. One of our neighbors filed for compensation and was paid a mere $8000. This, of course, did not cover his entire bill or cure the insect problems he will continue to fight in the future.

Though it is difficult to grasp exactly how severe insect resistance issues are becoming, a family friend, who happens to be a welder, has seen an influx of customers because they need him to reattach insecticide boxes to their field planters. They've noticed that their friends are doing the same. When Bt was first used, we thought that these chemical boxes would no longer be needed. The new insect resistant plants were supposed to defend themselves!

In addition to these unexpected chemical needs, Bt corn pollen can contaminate wild varieties via cross-pollination, one type of "gene-transfer" that is a concern to many farmers today. Because seed is light and can travel easily, cross-contamination can affect neighbors across the road, in the next state, and, even, as far away as the next country!

The unintended transfer of modified materials seems particularly worrisome as scientists begin to modify species that share reproductive material. This is the case with genetically modified canola, a crop used to make cooking oil and biodiesel. Canola, like broccoli, kale, cabbage, and brussel sprouts, belongs to the *brassica* family. It is not uncommon for one brassica plant to exchange genetic material with another producing a hybrid variation of brassica offspring. Many vegetable farmers worry that cross-pollinating GM varieties might have negative effects on brassica species elsewhere.[22]This type of gene modification threatens to unintentionally alter several food products at once.

As you can see, our abilities to control manipulated "transgenes" in real ecosystems remains limited. We have reduced complex, risk-inherent technologies to safe equivalents to satisfy immediate desires. In so doing, we have allowed powerful seed and chemical conglomerates to convince us that transgenes are necessary despite their long-lasting consequences.

According to a recent report called *GMO Myths and Truths*, these risk-inherent technologies may present even greater challenges than we previously thought. The authors explain how hundreds of genes may mutate unpredictably when a transgene is shot into a cell or linked to a bacterial sequence used to infect a cell. These processes create additional risks that genes important to some property could be damaged within our food supply.[6]

While there is no guarantee that such effects would be recognized in controlled research settings, such effects can and have become apparent in farm fields and surrounding ecosystems over time. Unfortunately, once the seeds have been sewn, the resulting disturbances are difficult, if not impossible, to remedy.[6]

In recent years, a few research teams have started to tease out why consumers often assume gene technologies are more precise and beneficial than we can actually prove. Researchers Latham and Wilson point out that such views are encouraged by politicians because they shift blame away from individuals and companies toward something that can be easily fixed: our genetic make-ups. This diversion reduces the pressure politicians feel to regulate and tax harmful products, actions that create tension within a politician's district. Latham and Wilson explain that by allocating tax dollars to biotech jobs instead, politicians can replace difficult decisions to tax and regulate with relatively easy decisions. Because they have found a way to simultaneously create new jobs and cure constituents' frustrations, the promotion of biotechnologies often meets widespread approval.[22]

While genome research has been a helpful tool to understand certain diseases, its potential usefulness has been drastically over-stated. For seemingly obvious reasons, most researchers are not as excited to discuss the limits of modern gene technologies as they are to discuss the seemingly limitless possibilities.

Nonetheless, the reality of the human condition is one in which both environmental and genetic factors shape human health. We typically have greater chances to improve our collective health by taking care of our surrounding environments than we do manipulating genes. As Latham and Wilson point out, the perceived promises of gene technologies even prevent some of us from undertaking less expensive and risky solutions to correct health problems.[22] These long-lasting solutions include alterations to our diets, stress levels, and exercise.

As we learn more about gene interactions, more scientists are waking up to these age-old remedies. New information is changing the way we view gene and chemical interactions. We now know, for example, that the majority of human ailments are not influenced by a single gene. Most diseases are influenced by numerous genes interacting in complex ways. One gene can code for several proteins.

We have also learned that what we originally described as "junk DNA"—the 95% of the human genome we thought had no function—actually plays important roles in the body. When you remove a small portion of this "junk," you can offset entire genetic sequences interrupting processes that make important enzymes and proteins. This "junk" DNA can serve a wide variety of functions including roles in regulating gene expression, evolution, and development.[23]

Recent research in the field of epigenetics, the study of our environment's influence on our genes, further highlights the limitations of our agbiotech knowledge. The question my generation was asked in grade school, *"Is an ailment due to one's genes or one's environment?"* no longer applies to many situations because it is too much of an over-simplification. The answer to this question is obviously both. Even people who do inherit genetic diseases are finding that they can manipulate and manage many disease symptoms via diet, exercise, and alternative therapies. We have learned that these environmental factors work to turn on and off portions of our genome.

While studying epigenetics, we have also learned that chemicals in the environment can influence the genes we pass on to our children! In a study on rats, scientists successfully detected the negative effects of an ag chemical called vinclozin over four generations. These effects did not subside over time.[24] An environmental toxin influenced the unborn because it left a permanent imprint on hereditary material. Who would've guessed?

Intuitively, we have known that this could happen for some time. Human eyes followed the results of Agent Orange and Hiroshima. We made just as many inferences about the heritability of chemical toxins during these times as those drawn from current studies. Now, however, scientists are beginning to better understand the mechanisms behind these inheritances.

One recently explored mechanism occurs when pieces of transgenic DNA act as heritable environmental toxins themselves. In some new types of GM plants, genes are silenced to change the type of RNA molecules produced by a plant. The RNA molecules produced then work to silence or inactivate gene expression during the modified organism's lifetime. Some new GM wheat and barley varieties, for example, utilize gene silencer technologies to change the type of starch made by a plant. Likewise, some varieties of GM insect resistant plants contain silencer genes that can act to kill pests feeding on a plant's altered molecules.[40]

Unfortunately, research has also shown that these altered molecules can be taken into the bodies of animals and humans eating transgenic plants. Gene silencing may then be inherited by these organisms' offspring. This potential risk factor is particularly problematic because altered RNA molecules are being developed to use across large swaths of crop land.[40]

Combine these risks with the fact that DNA from transgenic organisms can be taken up by free-living bacteria and it sounds like the human race has forgotten about evolution entirely. While examining bacteria in Chinese rivers, Dr. Ignacio Chapela found that some bacteria took up transgenic DNA strands and incorporated them into their genomes. The unintentional development of new diseases and bacterial strains could influence our abilities to fight pests and antibiotic resistance in the future.[41] All of this relatively new information has made us aware of additional risks associated with GM technologies.

While my father is aware of these risks, he does not know if they are influencing specific animal fertility measures. Nonetheless, he follows genetics research closely because he feels it's his responsibility to be alert to potential health problems when they do occur. Many of his colleagues, however, seem largely unaware of potential variations associated with GM technologies. They have already written such technologies off as unlikely contributors to animal fertility problems because these organisms have been deemed substantially equivalent.

Not only is it possible that the GM technology in feed crops is posing a risk to animal fertility, fertility issues may arise when genetically modified plants are treated with pesticides and fed to animals. While my father remains open to the idea that GM feed could play a role in causing fetal abortions and infertility issues on his client's farms, he is even more suspect of the chemicals applied to this modified feed. Specifically, he is worried about how glyphosate mixtures affect developmental processes within pregnant animals and animals lying on glyphosate treated bedding.

Genetic problems are likely exacerbated when feed chemicals disrupt the animal's gut bacteria affecting important metabolic processes. These bacteria would normally enhance digestion and help creatures defend themselves from the toxic residues found on animal feed and bedding products. New research exploring the effects of feed chemicals on fertility and cellular metabolism follows two decades of herbicide treated, feed consumption on the American farm and at the American dinner table.

In one North Dakota State University study, researches testing glyphosate formulations' effects on field peas found that only 2% of treated seeds germinated normally. Nighty-eight percent were abnormal missing essential structures or deformed to the point that they were considered non-viable.[25] Despite the State Seed Department's attempt to educate farmers that glyphosate formulations should *not* be used to dry down seed-crops, many farmers did not get the message. They continued to spray their crop with glyphosate to expedite harvest rather than waiting for it to dry naturally. These seed-crops include lentils, peas, wheat, and flax. These plants absorb the glyphosate, dry, and die, leaving their seeds ready for collection. Apparently, this method of harvesting also destroys the seeds' future viability.

These seeds then go on to feed other creatures, including humans. If you are a fan of non-organic split pea soup, you have probably consumed glyphosate treated seeds as the peas in your soup. Interestingly, while the research community is telling farmers to avoid using this chemical to dry down seed crops, they have occasionally advised farmers to feed glyphosate treated seeds to animals instead! Given recognized similarities between plant and animal nutrition and development, such advice seems extraordinarily careless. How can such grain be good for pregnant animals and their unborn if the seed itself is non-viable?

The effect of glyphosate on these seed crops is similar to many studies designed to evaluate the effects of low doses of Roundup herbicides on the reproductive health of animals. To date, scientists have reported that "glyphosate-based herbicides are toxic endocrine disruptors in human cell lines,"[26]and that they can induce cell death and damage human umbilical, embryonic, and placental cells.[27,28] They can also cause embryonic malformations in vertebrates,[29] including birth defects to frog and chicken embryos at doses lower than those used in agricultural fields.[30]

In one French study, researchers testing Roundup herbicides on mature male rats found that within 1 to 48 hours of exposure, testicular cells of the mature rats were either damaged or killed. Even at a concentration of 1 ppm, the Roundup decreased some test subjects' testosterone concentrations by as much as 35%.[31,32]Such effects are alarming given the similarities between our developmental mechanisms and those of the tested species.

What's more? These effects may be compounded by glyphosate's effects on bacteria. Because glyphosate's primary mode of action is via the shikimic pathway and this pathway is not present in mammals, most plant scientists and biotechnologists assume that glyphosate is not harmful to humans and animals. However, this same scientific community has often overlooked the fact that the shikimic pathway *is* present in many of the gut bacteria that help humans and other mammals digest their food. This year there is new evidence that glyphosate can interfere with the ability of the body's enzymes to detoxify foreign particles as well as its ability to form important amino acids. This includes an amino acid called methionine that the body must obtain from the diet. One group of independent scientists recently demonstrated that several diseases associated with the Western diet can be linked to a growing amount of glyphosate residues in our food and environment. These diseases include intestinal inflammation, depression, autism, infertility, cancer, obesity, and diabetes.[33]

These findings make a lot of sense to my father because he is convinced that rapid trait selection, genetic inbreeding, and other factors known to cause reproductive diseases cannot account for all of the infertility problems he's observed during the last twenty years. He suspects that either the glyphosate itself or some agent promoted by glyphosate interferes with normal hormonal development and reproduction.

In fact, at this time, he is working with a small group of researchers to better understand two substances associated with early embryonic death, vascular damage, and swollen placentas. These growths include a fungal toxin called zearalenone that is already associated with fertility problems (and glyphosate), as well as an unclassified agent the group associates with fetal abortions and swollen placentas in horses, swine, and cattle. Thus far, the group's findings are consistent with the observation that animals bedded on glyphosate treated corn stalks and eating glyphosate treated feeds suffer more severe long-term reproductive effects than their counterparts. They are also consistent with the observation that humans appear to suffer fewer reproductive issues of this nature. My father thinks that this is because glyphosate tolerances are typically much higher for animal products than human products.

Even if the group's suspicions are not confirmed, the other effects we've described certainly present fertility risks to both animals and humans.

The use of additional crop chemicals could exacerbate such risks. Seeds soaked in neonic insecticides, for example, are so damaging that a single corn kernal or wheat grain soaked in neonics can kill a songbird.[34] These insecticides certainly have the power to affect reproduction in the sense that, in some cases, they prevent it entirely. This means that both of our most widely used pesticides (glyphosate based herbicides and neonic insecticides) can negatively impact animal reproduction.

Animals, like plants, begin as a single cell. All of the genetic information that will later become an adult is concentrated in these tiny beings. In the presence of chemical toxins these beings *can and do* undergo changes that may influence every cell dividing off this origin.

In a GM soy-growing area of South America scientists reported that human birth defects and embryonic malformations, similar to those observed in other vertebrates, increased almost ten times during the same decade the region saw exponential increases in GM soy and glyphosate use. This report was based in a clinic that reviewed all of the region's children needing services for birth defects and embryonic problems.[35]

While such reports are devastating they should not be surprising. Pesticides have long been known to increase the risk of infertility and miscarriage by disrupting hormones and reducing sperm counts within human populations.[36-38] Our understandings of how such chemicals influence gene expression is advancing, but the basic relationship remains the same. Environmental pesticides can and do contribute to infertility by altering hormone levels in developing organisms.[38-39]

All of this is to say: we know much less about organisms' genomes, and how such genomes interact with chemical and environmental counterparts than is widely believed. There is no current study that will prove to a practicing veterinarian that his herd health issues are not, in some way, stemming from the chemically coated, GM crops he is feeding his cattle. Quite the contrary, there appear to be many risks associated with recombinant "crop protection" technologies like Bt and Ht systems.

Unfortunately, our failure to pay close attention to how modern biotechnologies are being used has left us in a predicament. We have set ourselves up to "scientifically support" the global distribution of technologies that can and already do cause unintended and irreversible modifications within global ecosystems.

In my childhood county, regular citizens use these technologies without thinking about them twice. A large billboard off Highway 13 is surrounded by fields of genetically engineered corn. In bold print it declares: **ABORTION: THE WORST FORM OF CHILD ABUSE**. I imagine the words **PESTICIDE TOLERANT CROPS** sitting in *abortion*'s place, followed by a string of other politically charged slogans: a child wearing a T-shirt that says: *DON'T GM MY FUTURE*, an image of a more agriculturally diverse Iowa next to: *GR CROPS: ENDANGERING MULTIPLE SPECIES*, a gene gun pointed at a double stranded helix: *DID GOD USE A GUN TO REARRANGE YOUR GENES?* a Mexican farmer holding up a genetically contaminated corn plant: *GM CROPS: AN UNINTENDED DESIGN,* and finally, *LIFE INC.* stamped across a child's forehead.

Where are all of the Bible Belt's right-to-life activists and scientific conservatives on the GM front? Why do I see so many anti-abortion billboards but zero billboards questioning current food production practices—practices that can potentially cause much more damage than the damage I associate with a single abortion? Sometimes I find myself wondering how many of my rural friends realize that crop-*protection* technologies are actually genetically manipulated life-forms!

Modifying heritable genes, especially to resist herbicides, can impact many people and creatures. It is difficult to discern who is responsible when genetic manipulations do cause unexpected consequences. There are no parents around to blame. Should the manufacturing company or the regulatory agency be held accountable? As we've learned in a variety of situations, manufacturers can eventually pay thousands to victims but no *one* person is typically held responsible. No one is there to make up for the health and environmental costs communities later deal with. Time and again, we've learned that the rights we extend in the name of corporate personhood don't live up to touted expectations.

When such atrocities happen at an international level, accountability becomes even more ambiguous. Should the mothers with deformed babies in Argentina sue the GM seed patent holder that bought up their land, the pesticide company associated with the seed, or their government? Do they even have the capacity to do so under international law or given their current economic situations? When someone is injured as a result of for-profit deceit, the guilty party usually gets off paying relatively little before moving on. How does one penalize such an entity in a way that triggers mindful reflection and change? More often than not, corporations are fined and shareholders go on leading relatively unaltered lives. Meanwhile, we grant more and more rights to this illusive corporate person!

The reasons for these complications are obvious. A corporation is *not* a person despite some people's confusion regarding this matter. A corporation does not feel guilt or regret in the sense that an individual feels these emotions. Trying such an entity for ethical wrong doings is difficult in any sector. This is why many argue that corporate influence over human rights and principles should be limited.

I do not currently think that it is wise to use gene transfer technologies in our food supply. Perhaps, someday, we will find less risk-inherent ways to incorporate GM technologies in feed crops and animals. As it stands now, sound scientific evidence supporting GM technologies is periodically interrupted by legitimate cries that these chemicals and technologies pose serious risks to our nation's youngest inhabitants and the environments they live in.

We, as a society, are far from answering difficult questions regarding when it might be acceptable to bioengineer an organism. Yet, legally, we have already approved the manipulation of genomes that evolved over billions of years. These decisions influence gene pools that have the potential to affect *all* life forms including their basic abilities to reproduce.

A good doctor knows when to admit that he or she doesn't have all of the answers. In this case, I'm pretty sure my father will never know exactly what's behind every fertility issue in his clients' herds. This fact, however, doesn't stop him from asking some very difficult and important questions so that he can attempt to remedy ongoing fertility problems.

DIAGNOSIS 5: SCHISTOSOMUS REFLEXUS

I've seen some strange and wonderful things on northeastern Iowa farms. None as eye opening, however, as the day I helped deliver a calf with a schistosomus reflexus. Ever since that day, I've been under the impression that it is a miracle life works at all.

Being asked to help pull a calf always presents exciting challenges. On a good day, you get to witness a newborn in its first moments. On a bad day, you realize just how fragile life really is.

After sleeving up, the vet hopes to coax the young creature out using simple maneuvers: by rotating the calf's body, untwisting its legs, and turning the calf so that it is more in line with the birth canal. With a little help, the pregnant cow should be able to push her baby out. She often accomplishes this entire process while standing and, with what appears to be, relatively little concern.

The process isn't necessarily pretty. The calf slips out through a film of blood and mucus as well as other bodily fluids and membranes. This is all followed by the rather large and slimy placenta. If you aren't disgusted yet, just wait until the farm dog decides to come over and eat the entire pile of afterbirth lying on the ground.

I try to focus on the mama-baby connection. Though there are some similarities to humans, there are also several immediate and remarkable differences. If she is healthy enough, the mama will often turn to her calf and lick him or her clean. This cleansing ritual is sometimes followed by an encouraging nudge, at which point the trembling newborn wobbles to her feet and stands. Imagine if a human baby were able to accomplish such a task! It takes us months just to realize we have feet meant for walking.

Then, as though this isn't impressive enough, the calf finds the teat and begins suckling. The mama can chew her cud and relax while her baby teaches itself how to nurse.

Learning how to nurse my son only took me three stress-filled days, followed by many weeks of frustration and uncertainty. The entire process was completely dependent on my ability to recognize my son's ambiguous hunger pangs, as well as an ample supply of patience as I directed him to the source over and over again. In the absence of a thoughtful, caring parent it amazes me that any human child can learn to eat! Yet, a baby calf will teach itself this entire process within just minutes of being born.

Pulling a calf is not always a wonder-filled experience. It can also be a veterinarian's greatest mental and physical workout. I've seen my father labor like a birthing woman—dripping sweat and straining against a cow's powerful muscles—standing in a mess of blood and manure. If the calf presents itself backwards, is deformed or its legs are twisted, it can often take quite some time to work it into a better birthing position. When this is not possible, it might need to be jacked out, or worse yet, cut out via c-section to save the mother. By "jacked out," I mean just that. The veterinarian anchors a jack to the calf's legs, sets the jack against the cow's derrière, and levers against her tailbone until the calf is able to slide through the birth canal. The cow often stands through this entire duration! If you're lucky, the calf is alive and healthy. Sometimes, you're not.

I had never seen my dad perform a fetotomy (cut a calf out) and didn't particularly want to. I understand that if a calf is dead and deformed and there is no other ways to retrieve it, you might need to do this to rescue the cow. That said, I had no desire to participate in this life-saving process.

So on that spring day, standing in the middle of a large pasture underneath an old oak tree, surrounded by an entire herd of on-looking cattle, I prayed that this would not be necessary. We had run into one of those storybook bad days: it was cloudy, it felt like rain, and a cow was dying before our very eyes. I had had my share of death for one day. Prior to this vet call, we had posted two dead pigs and euthanized a client's miserable, dying dog. We both hoped that things were about to look up.

Usually a client meets us prior to a delivery, but this particular client lived a few miles down the road. We happened to show up during chore time and he offered to meet us when he could. Without another pair of hands around, Dad started giving me more orders than usual. He sleeved up and I kept his tools sterile. He focused his attention on the birthing cow and I came up with inventive ways to keep the rest of the herd away from our work zone. After a lot of straining and cussing, I thought he would send me back to the truck for the jack. The work was almost over.

But he didn't ask for the jack. Instead, he told me to sleeve up. I grabbed a plastic sleeve and pulled it up to my shoulder. My adolescent arm swam in it.

"Your hands are smaller," he said. "I want you to feel this."

I was a little hesitant but more excited than anything. Hands-on work didn't usually encompass assisting with the actual delivery. I was simply the tool holder and retriever.

Dad seemed pretty confident that I might be able to feel something he couldn't—that I could maybe even reach the foot and turn it under. I applied some lubricant to my glove and, with proper coaching, eased my hand into the birth canal. I was taken aback by the strength of the muscles contracting against me. How could he find anything in this mess? And then, after blindly fumbling my way through, membranes began to take on more form. Almost immediately, my hand landed on a bulbous mass. Before Dad could give me the next directive, I was distracted. "This feels like a heart," I reported. "It's beating in my hand!"

"That's probably the placentome," he replied. "It feels like a button? Those things act like the umbilical cord in a human, transferring nutrients between the uterus and the calf."

"Why is it beating?" I wondered.

"It's not," he said, rather matter-of-factly.

"Um...yes it is!" I insisted. "I swear something is beating in my hand—just like a heart."

He was already exhausted. My commentary didn't help. "Let me see that," he grumbled. Frustrated, I watched my chance to assist end abruptly as he stepped forward to assess the situation.

"It's probably just pulsing with the contractions," he applied more lubricant to his glove, steered me out of the way, and started talking through the process again, moving his hand at different angles, searching and furrowing his brow. Finally, I got the bad news: it did feel like a heart! For some reason, there was a heart-like object thumping its way toward the birth canal. This was very strange indeed!

The laboring cow's legs began to buckle and dad became worried that she might give up and sit down. Everything becomes more difficult when a birthing cow sits down.

If things didn't start to improve, we would have to attempt a c-section, even though we couldn't ascertain if the calf was living or dead. Time was of the essence. We highly suspected that the deformed creature wouldn't live. This was, without a doubt, the most confusing vet call I'd ever been on.

A few minutes later we were on the jack. Dad had managed to work a deformed leg out but, oddly enough, couldn't find the other hind leg. It was a tricky situation which became even trickier when the cow decided she wasn't standing any longer and lurched to the ground. Dad's concern for the mother made him extra determined to get the calf out as quickly as possible. I crossed my fingers and held my breath: please let the mama live. Just let it live. And, then, suddenly, the calf was out: an amorphous mass of life, slumped against the ground: not breathing, not opening its eyes...not even twitching.

It was just lying there. This beastly thing: a freakish clump of tissues folded around bones and each other. I couldn't see any eyes, ears, or visible skin. "Is it dead?" I asked, not knowing exactly what to say.

Then we saw it: the beating heart. Sure enough, there it was: the dark, fist-sized maroonish-black mass I had accidentally held inside the cow's body. We stooped over and watched as it gradually slowed to a twitch, lifting attached arteries and connective tissue with each thump.

"It's a schistosomus reflexus," Dad admired. "Didn't know if I'd ever see anything like it."

"What's that?" I wondered.

"Essentially, a calf born inside-out."

After tending to the mother, we returned to the calf body, studying it intently. The poor creature died in less than an hour. Even so, I was amazed that it had lived at all—that it's heart had continued to beat through such a difficult existence and labor.

Dad pointed out that its skin was actually balled up in a tight tube where you would expect to find its spinal column. It's liver, stomach, and lungs sat openly visible to the naked eye.

I'd never seen anything like it and don't expect to again. This was life turned inside out in the truest, most bizarre sense.

A schistosomus reflexus is another condition that is difficult to pin down. Like every disease, it is influenced by a combination of one's environment and genetics.

Today's diseases, however, are not solely due to one's environment and genetics. They are also heavily influenced by political decisions. Many of the leaders who have shaped our FDA, EPA, and USDA regulations have got it all inside out and upside down. The American public must address an extreme case of schistosomus reflexus: this being our collective and often two-headed, agricultural misunderstandings.

An adult human consists of trillions of cells bound together in tissues, organs and systems, constantly talking to each other, texting this or that message to a neighbor or sending a chemical further down the wire. Cells in the brain can send messages all the way to the toe in a split-second. Chemicals produced by your cells can also make you feel happy, sad, depressed, and anxious. When these cells don't get the right messages, strange things start happening: tumors grow where they shouldn't, habits become difficult to break, and hearts develop outside the skin amongst other strange and perplexing conditions that plaque earth's organisms.

In addition to the DNA you were born with, we have learned that your body receives messages from the DNA of resident bacteria, fungi, and protozoans. Your DNA responds to both these residents and the foods and chemicals you ingest. These particles alter gene expression in ways that can be heritable or short-lived. Parts of your DNA can be activated and deactivated by where and how you live, by what you eat, and by how you think. The chemicals you ingest as a result of your food choices have a direct impact on your gene expression.

But there are also people making choices for you—regulating your genes for you. Today, without being made aware or given any choice in the matter, you are being exposed to insecticide neonicotinoids that can act on mammalian acetylcholine receptors, fungicide vinclozolins that can alter inherited stress responses, and herbicides that can induce DNA damage triggering early apoptosis, programmed cell death. [1-4]

During the last three decades, crops have gone from requiring zero genetic modifications, and a few pesticides and natural fertilizers, to crops that are often treated or genetically engineered to express several different insecticides and a fungicide before being planted and sprayed with additional herbicides, fungicides, insecticides, and more fertilizers. Some of these food items contain multiple toxins that may or may not be regulated as such. How safe is "safe" by current regulatory standards? Is "safe" safe enough to protect us from permanent damage to our own regulatory mechanisms: the DNA that directs our cells and the brain that coordinates cellular functions?

While our bodies are incredibly resilient, blast a DNA strand enough times with a gene gun or plane full of toxins and it has to bend, break, and adapt. We're only beginning to understand exactly how our bodies respond to all of these extra, synthetic chemicals in our environments.

Researches have known for several years, for example, that Parkinsons disease, the second most common human neurodegenerative disorder, is linked to chronic pesticide exposure. However, no one was able to explain how it might induce the actual disease. New research is shedding light on this issue.

My grandmother had Parkinsons. My family watched her gradually curl over like a question mark, no longer able to feed, clothe, and bathe herself. She lost her fine motor skills as the disease progressively damaged her dopaminergic neurons—cells in the central nervous system that help regulate controlled muscle action, mood, and stress.

Researchers now know that rotenone & paraquat, two agricultural pesticides, can induce mitochondrial damage and oxidative stress associated with the development of Parkinons Disease.[5,6]In addition, glyphosate formulations can trigger apoptosis (programmed cell death) and excessive autophagy of neurons to induce Parkinsons development.[1] An autophagic cell, like a PFTS pig, essentially digests itself.

This research is being published the same year that scientists have discovered that a common broad-spectrum insecticide, chlorpyrifos, can impact human brain development and the size of brain regions associated with social cognition, attention, language comprehension, sensory processing, motor coordination, and executive function. Research suggests that the EPA's current tolerance levels for chlorpyrifos may not be strong enough to protect developing children. While the EPA banned this chemical from household use in 2001, agricultural use is still permitted and common.[4]

In addition to these findings, scientists studying neonicotinoids, the insecticides that are so harmful to bees and birds, found that they can negatively affect human neurotransmission. These chemicals can bind permanently to nicotinic acetylcholine receptors in the brain. Research has demonstrated that invertebrate numbers and birds dependent on insects have declined in areas where neonics are used.[7,8]

Each of these pesticides has previously been shown to affect other animals. We should not be surprised then that they pose serious threats to human beings. Prenatal and childhood exposure to pesticides are increasingly identified as risk factors for nuero-disorders including ADD, autism, bipolar disorder, and learning disabilities. Again, this is rather unsurprising given that one of our supposedly safest herbicides can trigger cell death and self-digestion.

Thus far in this book, I have focused on one chemical cocktail more than others, but, as you can see, it is certainly not the only dangerous chemical on the market. I was interested in glyphosate formulations, like Roundup, because they are used more than any other herbicide on our planet and because glyphosate is currently under review by our

Environmental Protection Agency (EPA). If all goes well for the manufacturer, glyphosate will soon be re-registered. This means that different formulations containing glyphosate can continue to be sold legally in the United States. The current pesticide review process for glyphosate, scheduled to occur once every 15 years, may last six years from start to finish.[9-11]

Given glyphosate's notoriety and my father's concerns, I was rather curious about the ways our government regulates ingredients and contaminants in our food supply. How does a regulatory agency continue to justify the use of glyphosate formulations, on such a grand scale, when they have been linked to so many environmental, animal, and human health problems? Furthermore, how does the regulation of these herbicide formulations compare to the regulation of other crop chemicals on the market? While exploring these questions, I learned that our pesticide regulatory system is a complex, twisted up mess, in and of itself.

Pesticides regulated by the U.S. EPA include insecticides, herbicides, fungicides, rodenticides, acaricides (mite-killers), prions (folded-up proteins), and microorganisms.[9] Before a chemical manufacturer can legally sell and distribute these pesticide products, it must obtain a permit from the EPA. To evaluate a pesticide's safety, the EPA must review at least 100 different scientific studies submitted by the applicant, typically the chemical manufacturer. Any member of the public may also submit comments during open comment periods.[9]

While I am certain that there are many skilled and intelligent scientists conducting pesticide reviews for the EPA, there are too many flaws built into the review process to offer reassurance that these reviews are keeping us safe from harmful toxins.

Unfortunately, some of the problems are unavoidable. The indirect consequences of any pesticide can be overlooked in lab settings. Furthermore, controlled field tests are often difficult to conduct and fund. Regulators must draw their own conclusions from the studies submitted. This is nothing new. Laboratory studies are, by nature, over-simplifications.

We also mustn't forget that the nature of these studies is often rather complicated. Data might reveal, for example, that test rats are less sensitive to specific compounds than other animals or that it is too difficult to evaluate the risks of a chemical that disperses easily in an environment. For classification purposes, the EPA bases its recommendations on the results obtained, but again, these results may not be representative of what actually occurs in real environments.

While these problems are difficult to avoid, other oversimplifications *are* avoidable. For example, we currently fail to conduct thorough investigations of each chemical formulation marketed. With one sweep of the pen, the EPA can approve hundreds of formulations because the data used to evaluate pesticides is based on the active ingredient alone. This is despite the fact that a chemical like glyphosate is rarely applied by itself to crops. These glyphosate formulations often contain different chemicals with their own toxicity thresholds, but are still treated as one and the same by the EPA.

When the EPA last evaluated glyphosate, 56 different products contained the active ingredient. Today, our market distributes hundreds of different glyphosate-containing products.[12,13]Let me remind you that this is just one chemical in the United States we're talking about. Nevermind that all of the pesticides regulated by the EPA are inherently toxic. Each needs to be thoroughly reviewed to protect consumers from unnecessary harm.

While the EPA seems fully aware of this limitation, it continues to approve these products anyway. In the EPA's Ecological Risk document, the authors admit that there is little to no data on several formulations. This is concerning because some formulations are known to contain chemicals that may be more toxic than the active ingredient itself![12,14]Together, these ingredients may create even more damaging, synergistic effects to humans, animals, and the environment.

Some of these other ingredients, called "adjuvants," include surfactants. A surfactant can help spread the herbicide across plant leaves to enhance chemical uptake. One surfactant, called POEA (polyoxyethelene tallowamine), can cause more severe respiratory effects and lung damage to test rats than glyphosate does on its own.[15] A recent French study evaluating the toxic effects of the POE-15 family, which includes POEA, concluded that these adjuvants are even more toxic to human cells than glyphosate.[16,17]Other studies have demonstrated that POE-15 chemicals are thought to play a major role in glyphosate's toxicity to aquatic organisms.[18]

Remarkably, these adjuvants are not always included on product labels. Glyphosate and manufactured formulations like Roundup are treated as one and the same, even when such formulations are obviously more dangerous than the active ingredient. The potential hazards included in each of these untested, unregulated mixtures are then sprayed across millions of acres of American land and water! The EPA should demand that information about each ingredient is provided so it can better regulate *each* formulation used. It should also treat these adjuvants as active ingredients rather than misleading the public to believe that products like Roundup only contain one active ingredient in need of regulation.

In addition to granting permits, the EPA establishes chemical tolerance levels for use in food and animal feed products. These tolerance levels let consumers know what dilution rates are required to avoid a chemical's toxic effects. Tolerance levels often vary for different products. While the EPA's current tolerance level is 20 ppm (parts per million) in soybeans, it is 13 ppm in animal feed, 3.5 ppm in sweet corn, and .1 ppm in poultry.

Interestingly, several of the levels at which researchers have found glyphosate-based herbicides to be toxic to human and animal cells are much lower than the EPA's recommended tolerance guidelines.[12,19]

Numerous laboratory tests have shown that glyphosate-based

herbicides can cause genotypic damage and alter endocrine functions at levels far below established tolerances. At rates as low as .5 ppm researchers have observed endocrine disruption in human cells. Between 1 and 10 ppm, glyphosate-based formulations have damaged DNA, triggered cell death, suppressed mitochondrial and enzyme functions, and caused placenta and embryo damage. What's more? Both POEA *and* AMPA, glyphosate's main break-down products, contribute additional toxicities to these formulations.[19-21]

The effects demonstrated in these labs could explain serious problems with cell differentiation, liver metabolism, reproduction, behavior, and development. These effects have been observed in a wide variety of test animal populations over the last two decades.[19-21]

Interestingly, the EPA has granted increased tolerances to chemical manufacturers during several product reviews. These tolerance adjustments appear to be tagged onto a review process in much the same way our politicians tag riders onto bills. The differences between initial tolerance levels and current levels are displayed in the following table:

Increased Tolerance Levels Granted to Monsanto Since Glyphosate's Introduction		
Crop	***Initial Level (ppm)***	***Raised Level (ppm)***
Soybeans	6	20
Sugar beets (whole)	.2	10
Sugar beet roots	10	25
Alfalfa (forage)	75	175
Alfalfa (hay)	200	400
Data adapted from pgs 45-46 CFS comments to USDA APHIS "Herbicide-Resistant Crops and Weeds" as included in appendix to Glyphosate Review Docket (June 2009)[12]		

Note that this table only includes those tolerances granted to Monsanto and does not include tolerances granted to other chemical manufacturers. In one case, the EPA granted Dupont a significantly increased tolerance level for grain fractions fed to beef cattle. The acceptable tolerance limit for glyphosate in this feed product went from 200 ppm to 310 ppm![12] Observed toxic effects of glyphosate based herbicides have occurred at levels many times below these values.

Of course, tolerance levels are meant to serve as guidelines. They do not necessarily protect you from the millions of foreign molecules entering your systems, even when they are met. However, when these tolerances are inaccurate or unknown, as is the case with several widely distributed herbicides, the risk that some unknown toxic damage may occur is even greater than usual.

Even when there is only one active ingredient, problems can arise. Take atrazine for example. Atrazine, a toxic, agricultural herbicide, is applied before and after planting to kill weeds and has a tolerance level of just 3 ppb (parts per billion) for public drinking water. Even at this very low concentration, a glass of water containing 3 ppb atrazine can contain over a quadrillion molecules of this pesticide.[22] While "parts per billion" and "million" sound incredibly small, a considerable number of particles are present at the molecular level. These threshold values cannot tell us exactly how each atrazine molecule will interact with body tissues. How many molecules does it take to disrupt human development? For one man-made molecule to take the place of another at a crucial stage in one's life?

In 2007 alone, the world consumed roughly 5.4 billion pounds of pesticides.[23] This is almost a pound of chemical toxins for every man, woman, and child on the planet! What's more? This number doesn't even include non-pesticidal toxins! Given the fact that many of these toxins can cause permanent damage in developing organisms, it's amazing that any child can think straight or grow up to birth healthy children.

According to the Food Quality Protection Act (FQPA) passed by congress in 1996, the EPA will work to *ensure with a reasonable certainty that no harm will result from the legal uses of the pesticide and build an additional 10-fold safety factor into risk assessments to ensure the protection of infants and children, unless it is determined that a lesser margin of safety will be safe for infants and children.*[24] Every indication leads me to believe that the EPA's intentions are good.

From what I can see, however, neither of these guidelines is adequately addressed. In addition to the flaws discussed, a herbicide like glyphosate is regulated strictly on a per dose basis, even though it is, in many cases, used millions of times more than other regulated chemicals. By ignoring the quantities currently used, the resulting recommendations do little to protect infants, children, and adults.

Meanwhile, the mandate to protect against *unreasonable* effects continues to be ignored. Several assessment areas seem particularly lacking. In one Human Health Risk assessment, the EPA decided to waive the FQPA's 10X safety factor for infants and children stating that a 1x safety factor was sufficient. An organization called Beyond Pesticides pointed out that "this decision is flawed in light of the science" citing several studies documenting formulations' effects on embryonic development and hormone disruption.[25] One must wonder why there would ever be a need to waive this standard. If the product is truly safe, it should live up to each standard set.

Of course, there are no limitations on a manufacturer's say in a toxin's approval. While the product manufacturer should be expected to provide the EPA with a thorough list of studies, as it currently does, the EPA should also work to ensure that at least 50% of the research used to evaluate products has no direct association with the manufacturer. Conflicting interests lead to conflicting science. Possibly this is one of the reasons why the EPA's human health assessments historically cite so many studies that positively describe glyphosate products. Meanwhile, references to studies describing glyphosate's long-term effects on the endocrine and nervous systems have often been missing from regulatory resource lists.

The EPA isn't the only agency under public scrutiny for the way man-made chemicals are incorporated into modern food products. In addition to the EPA's pesticide regulations, the USDA regulates gene manipulation in agricultural crops. A division within the Animal and Health Inspection Service (APHIS) called Biotechnology Regulatory Services (BRS) has been given the authority to regulate and "facilitate" animal and plant biotech projects. The BRS has a mission to "ensure the safe importation, interstate movement, and environmental release of GE organisms."[26]

In response to criticism, APHIS recently announced "improved" regulatory processes including an intention to work closely with the EPA to review the "potential human health, plant health, environmental, and other relevant impacts" of genetically modified crops.[26,27] Like the EPA's improved processes, these improvements appear to do more to streamline overwhelmed systems than to protect consumers, plants, and the environment.

The Biotech Regulatory Service's primary means of communicating with the EPA is a document called the Environmental Impact Statement (EIP). Like the EPA's re-registration review, the EIP document is posted during a review for public comment.

Regulating Agencies Overseeing GM Plant Products

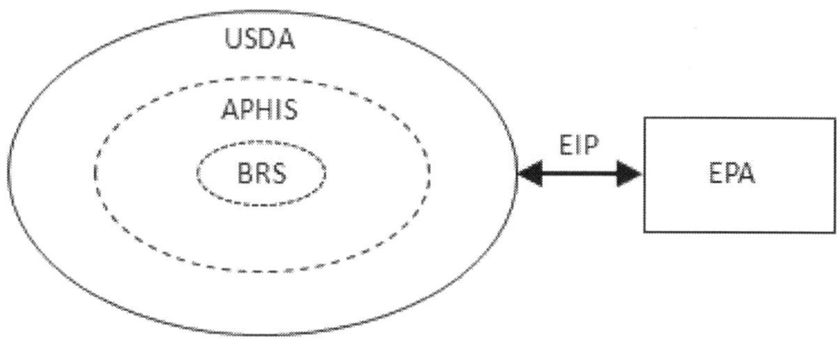

Once the GM product has been reviewed, the BRS can decide to issue a license to the seed manufacturer as long as the GMO does not pose risks as a plant pest. Unfortunately, experts often disagree as to which products pose as plant pests. In June 2012 the BRS moved toward deregulating three crops: 2,4-D and Glufosinate Tolerant Soybeans (Dow), Glyphosate and Isoxaflutole Tolerant Soybeans (Bayer) and Rootworm-Resistant Corn (Syngenta).[27] Each of these products contains new genes that are not prevalent in natural crop ecosystems. As we discussed in the last chapter, these new genes may alter gene expression via gene flow mechanisms. There is also the potential for weeds and insects to develop resistance to genetically modified varieties. When present, these noxious, resistant weeds and insects present a much greater burden for farmers to kill. These noxious, resistant weeds, created by our dependence on GM technologies, can certainly be thought of as pests!

As of July 2012, APHIS had deregulated 91 "transformation events or lines," several of which contain multiple transgenes. Among these events, are genes modified to resist the chemicals glyphosate, glufosinate, phosphinothricin, and imidozolinone. Crops have also been modified to be drought resistant, lepidopteran resistant, insect resistant, corn rootworm resistant, corn borer resistant, and virus resistant. In addition, fruit ripening, omega 3 fatty acid, flower color, and a variety of other crop characteristics have been *improved* for commercial markets, including transgenes encoding for plant fertility.[28]At this rate, we may have *improved* transgenic events for every marketable fruit and vegetable characteristic within a few decades!

In addition to increasing the number of crops we modify, American farmers are increasingly reliant on seeds that contain more than one modified gene. These products are often referred to as "double and triple stacked" traits. During July 2012, 20 product petitions were opened for deregulation status.[28] Eight of these included a glyphosate resistant transgene and nine were modified to express double or triple stacked traits. When approved, these premixed plants are doused with premixed chemicals. The following chart is provided to give readers a snapshot of GM transgene products pending USDA approval during the time period discussed.

GM Crop Events Pending Commercial Approval by APHIS - July 2012		
Company Requesting Deregulation	**Regulated Crop**	**Transgenic Event(s)**
Dow AgroSciences	Soybean	2,4-D, **Glyphosate** and Glufosinate Tolerant
Dow AgroSciences	Soybean	2,4-D and Glufosinate Tolerant
Bayer Crop Science	Soybean	**Glyphosate** and Isoxaflutole Tolerant
Dow AgroSciences	Corn	2,4-D and ACCase-Inhibitor Tolerant
Bayer	Cotton	Glufosinate Tolerant, Lepidopteran Resistant
Pioneer	Corn	Glufosinate Tolerant, Lepidopteran and Corn Rootworm Resistant
Monsanto	Soybean	Dicamba Tolerant
Monsanto	Soybean	Increased Yield
Monsanto	Canola	**Glyphosate** Tolerant
Monsanto	Corn	Tissue-selective **Glyphosate** Tolerant
J.R. Simplot	Potato	Low Acrylamide Potential and Reduced Black Spot Bruise
Pioneer	Canola	**Glyphosate** Tolerant
ArborGen	Eucalyptus	Freeze Tolerant and Fertility Altered
Syngenta	Corn	Corn Rootworm Resistant
Okanagan Specialty Fruits	Apple	Non-browning
Genective	Corn	**Glyphosate** Tolerant
Stine Seed	Corn	**Glyphosate** Tolerant
BASF Plant Science, LLC	Soybean	Imidazolinone Tolerant
Monsanto & Scotts	Creeping bentgrass	**Glyphosate** Tolerant
Source: data obtained from USDA Petitions for Non-regulated Status Pending, as of July 24, 2012 at: http://www.aphis.usda.gov/biotechnology/not_reg.html		

Various groups of consumers, environmental groups, and farmers have joined forces to oppose some of Dow's newest corn products, especially those containing both glyphosate and 2,4-D, an ingredient in Agent Orange, the Vietnam War defoliant.[29] These GM crops are being promoted as one of several solutions to glyphosate resistant weeds. If the glyphosate doesn't kill these plants, the 2,4-D will surely get the job done!

Unfortunately, 2,4-D has already been linked to endocrine disruption and reproductive toxicity as well as non-Hodgkin's lymphoma and liver damage. In a letter to the EPA during an open comment review of Dow's 2,4-D resistant corn, 70 physicians, nurses, and public health specialists worked together to urge the EPA to deny Dow's request for product approval. They cited various reasons why 2,4-D is not a wise choice for American farmers. They also expressed concerns that human exposure to 2,4-D is already too prevalent and that the chemical can disperse its way into people's homes. According to this letter, children living downwind of farms already come into contact with 2,4-D molecules distributed by the wind.[30]

If 2,4-D resistant crops do become more common, their use could mimic trends we've seen with glyphosate.[29] This 2,4-D dependent system will present many of the same problems we've already observed with Roundup Ready systems including chemical dependencies and weed resistance. Despite the implications for developing organisms, we continue to jump back on our toxin-charged treadmills. This type of solution to our pest and weed problems hardly seems acceptable.

The regulatory decisions used to propel ag pesticides and genetically modified seeds into the market, are anything but beneficial for many. These decisions are upheld at a time when scientists from around the world are trying to figure out what is happening to our honeybees, bats, ladybugs, frogs, and grasshoppers. Yet we continue to poison the insects and pollinators that make food possible, not to mention the predators that keep insect pests, like soybean aphids, in check.

Many insects, like honeybees, rely on field crops for food. When their food is sprayed with toxins it becomes an ingested poison. If you are a pollinating insect attempting to survive in this environment, you need to gather your food and water from plants that have been treated with numerous pesticides including neonics (like clothianidin and thamethoxam). Insects relying on corn pollen as part of their diet are dealing with poisons that occupy a surprising portion of arable land. Scientists estimate that 90% of America's 92 million corn acres are treated with neonics each year.[7,8]

Like other hazardous chemicals that have gone before it, neonic and glyphosate formulations have escaped widespread criticism for decades. The same regulatory agencies that approved dangerous products like DDT, PCBs, and BPA continue to approve chemical dependent, GM cropping systems.

While our regulatory agencies must grapple with bureaucratic and scientific challenges, chemical companies remind consumers that there are way too many variables to ascertain whether a specific chemical is the source of a problem.[31,32]I would have to argue that the lack of specific information combined with a link to so many problems seems telling in and of itself.

A schistosomus reflexus is unlike many diseases because we really know very little about how it begins. Are the developmental abnormalities a product of the mother's diet, a genetic mutation, genetic inbreeding, or an environmental toxin? Can we have any influence over them to reduce their frequency? We deal with the problem as best we can than move on, because no amount of scientific research is going to explain away these problems, much like the miracle of birth that brought this calf into the world in the first place.

When a cow accidentally swallows a piece of hardware, on the other hand, say a nail, staple, or bit of bailing wire, there is a recognizable treatment that will cure the problem. A veterinarian obtains a large magnet, roughly the size of a twinkie, sticks a tube down the cow's throat, and has her swallow it. With a bit of luck, the sharp object will cling to the magnet preventing the development of harmful lesions and abscesses that might permanently damage the cow's digestive system.

We need to eliminate the variables that are preventing us from fixing regulatory problems. While no regulatory system is perfect, we can make several changes that will prevent unnecessary harm and damage to our nation's inhabitants. If we continue on our current trajectory, the new and innovative technologies designed to improve our lives will only turn diagnosis and treatment into a giant inside out, upside down mess. These poorly regulated chemicals put our entire interdependent web of supporting life in danger.

DIAGNOSIS 6: DISPLACED ABOMASA & CLOSTRIDIA

Middle school students are always very curious about animals that eventually become food. When my students find out that my family raised beef cattle, their questions about cows could go on for hours. In addition to questions that lead to productive conversations, a few attention-seekers wonder if I've ever been "cow tipping." They've seen "cow-tipping" escapades in comical movies and want to know if this is really something rural folk do to entertain themselves.

To be honest, I've always been somewhat perplexed by the whole notion of cow tipping. Could someone really tip a cow over if they wanted to? What is it about a cow, rather than a horse, that makes it the ideal animal for this on-screen prank?

The students asking these questions likely assume that cattle are dim-witted creatures that sleep standing up, as they appear to on film, and that they are easier to tip over than other creatures their size. I take this opportunity to explain to my students that cows, in fact, do not sleep standing up and are actually very easy to startle. To top this off, there is no way that one human being can push over a half-ton animal without doing it harm. Silly movies! Silly kids!

Nonetheless, I am glad I get asked this question, because it's a great segue into some interesting conversations about cow-rolling, a perhaps equally intriguing phenomenon that is actually related to my middle school science curriculum.

While most farm kids don't grow up rolling cows, I have helped roll cows on various occasions to treat a condition called a displaced abomasum or D.A. If a farmer called our house to ask for support with a D.A, my dad was often interested in some familial assistance.

A displaced abomasum or twisted gut, as some farmers call it, occurs when a cow's true stomach (abomasum) becomes filled with gas causing it to rise to the top of a cow's abdomen. Sometimes it becomes twisted as it rises before settling in on the cow's upper left side.

This condition is concerning because it can seriously decrease a cow's productivity and lead to numerous complications. In order to treat it, a veterinarian must move the cow's abomasum back into its normal position. While surgery can be performed, it often isn't necessary. More often, a D.A. can be treated by rolling a cow onto its back. This allows the less dense abomasum to drift upward before the vet fixes it to the abdomen with a few stitches. By stitching it in place the veterinarian can often prevent a relapse, a common issue when a cow has a D.A.

Rolling a cow is no small feat. Usually two to three helpers are required to make sure the half-ton creature doesn't kick anyone in the process. To prevent injury, Dad binds the cow's hind legs before pulling the rope over a gate or barn sill for extra leverage. Cow's have extremely strong legs, a fact my father knows all too well after having his knee crushed when he was kicked by a cow in 1974, his first year of practice.

Most D.A.'s occur soon after a cow has a calf. During pregnancy, the uterus puts pressure on the abomasum moving it upward. After giving birth, the abomasum should eventually return to its normal position. If indigestion or poor nutrition alter the cow's stomach contents, gas can accumulate and carry the cow's entire stomach upward like a hot air-balloon.

When diagnosing a D.A., the veterinarian can actually hear a "pinging" sound over the area of displacement. Distension from the accumulating gas causes the skin to resonate in an unusual way when you snap your fingers against atypically taut skin. Using this sound, the veterinarian can decide which way the cow should be rolled and where, approximately, the stitch should be placed to hold the abomasum down.

As unusual as this may sound, D.A.'s aren't the only problems that arise when gas becomes trapped in a cow's stomach. Any change in normal gas production can cause bloat. When a diet contains too much starch and too little fiber, gas becomes trapped in the rumen and it inflates, pushing against the cow's other organs and lungs. Only by relieving the tension can the bloated cow return to a normal, productive, and comfortable state.

If bloat is severe enough, a veterinarian may need to fistulate the cow with a bloat needle, or cannulae. Using a hollow needle, the veterinarian can release trapped gas from the cow's rumen. If the cow is no longer able to release gas itself, the port can be made into a permanent access point in the cow's side, similar to the ports left in chemo patients when they receive regular doses of chemotherapy. This port can be opened and closed regularly to monitor gases, body fluids, and sources of possible infection.

Bloat, like a D.A., is usually a result of inadequate nutrition. In both conditions, excess gas accumulates due to bacterial imbalances in the gut. To understand what's going on exactly, one needs to have a *rumen*entary knowledge of a cow's gut. Understanding the rumen, in turn, requires an appreciation of bacteria, the tiny creatures that allow the rumen to break down foods into digestible nutrients and minerals.

Inevitably, as a public school teacher, I am gifted at least three containers of antibacterial hand soap every year. Some teachers buy it in bulk dispensers and set them on their desks for regular student use. Students concerned about bacteria are often already supplied. They like to remind me that they don't need to wash their hands at the lab sink because they can use their pocket-sized containers of antibiotic hand soap after messy labs.

To be fair, a middle school is probably the ideal place to create a germophobe, so I can understand an appreciation for such products. Some middle schoolers haven't quite picked up on the whole nose-blowing phenomenon, preferring to wipe it where they will. This observation, combined with the fact that 120 students travel through the same confined space, commingling the germs they've brought from God Knows Where, does put one on edge when you know a major cold or bug is going around. Like most of the staff at my school, I plan on catching a cold after every major vacation and hope that it's not as bad as the last post-vacation epidemic.

Students who suffered through the media's over-blown swine flu scare in 2009 were incredibly concerned about germs. We took measures as a staff to reduce the spread of contagious bugs by frequently reminding students to keep their hands to themselves and to wash up after interactive activities. We were also provided with disinfectant cleaners and told to wipe down student desks as often as possible.

Some students are already concerned about bacteria anyway. This may stem, in part, from an increasing number of students who appear to have chronic allergies to things as common as grass. Two years ago, I had to curtail outdoor labs because at least three students in every class harbored serious allergies to pollen, dust, grass, or bees. Kids with chronic allergies tend to be the same students who aren't used to getting dirty or putting their hands on things because these things threaten their very wellbeing. Students also hear all kinds of ridiculous garbage about bacteria on television. We've all seen the commercials that have demonized germs to help market a product.

Imagine their surprise then when we begin our unit on biotic organisms and learn that bacteria are needed to keep the human body healthy—that our bodies are teeming with thousands of microbes that are metabolically active, working in our noses, mouths, and skin to produce inflammation-fighting chemicals, and in our guts to help us digest proteins and fats. In fact, scientists refer to the trillions of bacteria coexisting within us as members of our *microbiomes*. Each one of us is a breathing, walking biome of life!

Between one and three percent of a healthy human's body weight is made up of bacteria. This means that a one-hundred pound middle schooler could easily harbor three pounds of bacteria! Each cell can hold between five and ten bacteria not including the organelles that evolved from bacteria originally. So in addition to the 20,000 genes each of our cells normally has, cells have additional DNA in resident bugs. In fact, microbes add millions of genes to the human biome!

What's more? Even healthy humans harbor the bacteria we typically try to avoid. When a person is healthy, he or she can keep these harmful bugs in check. The good bacteria in our bodies and soils control and balance the bad.

We have biased interests toward our own microbiomes, but bacteria play even more important roles in other species. Sometimes these microbiomes extend beyond what we typically observe to be a single living organism. The soybean, for example, depends on nitrogen-fixing bacteria. These bacteria harvest nitrogen from the air and soil and deliver it to the plant. The bacteria, in turn, appreciate a cozy habitat on soy root nodules. The plant will later draw the bacteria's "fixed" nitrogen up through the roots and use it to build important chemicals like hormones and proteins.

Earth's 150 species of ruminants, including cattle, host a particularly large quantity of bacteria. Grass-eating goats, sheep, elk, giraffe, and bison rely on these creatures to help digest plants and grains. Because these vertebrates lack the enzyme cellulase, none are able to digest cellulose-containing plant matter on their own. Ruminant herbivores must completely depend on the bacteria, protozoa, yeasts, and fungi living in their stomachs to help them digest food.

Most of these resident microorganisms are concentrated in the rumen, one chamber in a cow's four-chambered stomach. No, for the thousandth time, cows do not have four stomachs! I get this question *a lot*. Scientists estimate that 1 ml of rumen juice can contain up to 50 billion bacteria! This is a much more densely and diversely populated microbiome than our own.

The four parts of a cow's stomach are the rumen, reticulum, omasum, and abomasum. In the first two chambers, the rumen and reticulum, the food mixes with saliva and separates into layers. Solid layers clump together to form a cow's cud. The cud is then regurgitated, chewed up, mixed with saliva and voila: we have a tasty dessert: particle-sized bits of plant fiber that are swallowed again to reenter the rumen. Bacteria begin their important work here, fermenting the starch-filled plant bits, so they can be turned into more usable forms like sugars and fatty acids. Fermentation continues as this material moves through the first two chambers and into the omasum where water and minerals are absorbed by surrounding blood vessels.

Finally, the food is ready to enter the cow's "true stomach," the abomasum, which functions very similarly to a human or pig's stomach. Enzymes in the abomasum degrade food particles further. This food is then passed on into the small intestine where most nutrient absorption occurs.

Because a cow's gut lacks oxygen, anaerobic bacteria (bacteria that thrive in low-oxygen conditions) break down food particles via fermentation. In addition to producing usable sugars and fatty acids, bacterial fermentation produces large amounts of acid and byproduct gases like carbon dioxide and methane. As these bacteria work, they also make vitamins, detoxify plant poisons, recycle compounds, and build new proteins.

The relationship between the cow and its bacterial residents is mutually beneficial. The cow provides the bacteria with living space and food. The microbes, in turn, convert available plant matter into the cow's main source of energy while carrying out processes that keep the cow healthy. If the bacterial content of a cow's gut is altered, it can become a serious threat to the cow's health.

Because these bacteria essentially build and break down the materials used by a cow, alterations in their content can also influence the nutritional content and over-all health of the milk and meat we derive from the millions of cows living on our planet.

Like humans, cows normally harbor an array of helpful and potentially harmful bacteria. Bacterial imbalances can lead to various illnesses. Sometimes these same bacteria create conditions in which the cow is unable to absorb available nutrients from its food. As a result, a cow's gut can become extra acidic, irritated, and inflamed. Corn, which is less fibrous and more acidic than grass, can also disrupt the normal pH of a cow's rumen causing acidosis. A cow with acidosis not only feels sick, it may stop eating or "go off feed." If untreated, the cow can develop diarrhea, ulcers, and bloat. These conditions further weaken the cow's overall immune system. To a farmer, these issues are bad for production; to the cow these issues are super uncomfortable.

Even when a cow is healthy, the bacteria in its 40 gallon rumen create an abundant amount of gas. This gas is eliminated through regularly farting and belching. When bacterial imbalances disrupt digestion, farting and belching increase. Today farting and belching ruminants create so much gas that several researchers have investigated ways to reduce ruminant emissions. In fact, some experts have proposed that we tax giant manure pits where methane accumulates over fermenting cow poop. In the United States, a small percentage of dairy farmers recycle this methane by harnessing it and converting it into electricity.

Other researchers have looked into ways we might modify cows' diets. Because diet affects gas production, simple dietary changes can significantly reduce bovine methane emissions.[1] Adjustments to animal feed intake can also reduce bacterial issues that lead to bloat and displaced abomasums. This shouldn't come as a surprise given what we already know about the differences between the diet cows evolved to digest and the diet many cows are now fed.

Traditionally, most cows ate grass. Their rumens are designed to process forage, turning fibrous cellulose in grass into proteins and fats. Today, however, most American cows are fattened up in commercial feedlots where they are fed mixed rations of grain, primarily corn, supplemented with antibiotics and growth hormones. Occasionally, as a means to cut feed costs, these cows are fed industrial byproducts.

Byproducts include a variety of ingredients. I've heard of all kinds of things including potato chips and chocolate chips! One of the most commonly purchased byproducts in the Midwest is a corn-distiller diet, a product of the ethanol manufacturing process. By attentively managing the modern cow's diet and supplementing it with antibiotics and growth hormones, an industrial producer can now take a beef animal from birth to slaughter (90 pounds to about 1200 pounds) in a year and a half! A century ago, it took four to five years to achieve similar weight gains in grass-fed beef cattle.

Rapid weight gain is one of several efficiency measures industrial ag proponents like to boast about in ag publications. Recently, consumers have become concerned that regular antibiotic use in animal feeds and injections (to boost productivity and strengthen animals' immune responses) is creating new strains of antibiotic resistant bacteria. They point out that much of the stress that drives routine antibiotic use is only the result of the weird diets we feed them.

In fact, veterinarians often advise farmers to turn their cattle out to pasture when their guts become overwhelmed by bacteria that increase the incidence of bloat and displaced abomasa. Here they can feed on grasses and legumes. These are the same grasses and legumes cows' ancestors evolved to eat over countless generations. Not surprisingly, the veterinarian's miracle cure is simply the cow's original diet!

While industry wastes money attempting to prove that there is no direct correlation between feedlot diet, bloat, and acidosis, several veterinarians, forage specialists, and beef specialists already recommend diet-dependent prescriptions for these problems. By advising clients to turn their cows out to pasture, these experts suggest that increased fiber in-take can reduce the occurrence of these problems.

Of course, grass-fed cattle are not immune to bloat. However, according to one bloat prevention and treatment guide, "proper management can reduce or eliminate bloat problems as effectively as purchased supplements." These proper management techniques include providing a consistent diet, controlling access to plants that cause bloat, and utilizing rotational grazing practices when possible. By rotating the cow from field to field, a farmer can provide more uniform forage quality and reduce the incidences of bloat.[2]

Typically, cows in a feedlot don't have access to these conditions. In fact, many veterinarians and specialists treat "feedlot bloat" and "pasture bloat" as separate conditions because they are so dependent on the cow's environment & diet. Feedlot bloat occurs when cattle are fed high-grain rations that may or may not contain enough forage materials. In addition to a lack of forage, cattle consuming feedlot diets may experience bloat when the grain portion of the diet is too finely ground. As it turns out, the saliva stimulated when a cow chews its cud actually aids digestion. When the grains are already broken down into finely manufactured bits, rapid digestion is accompanied by rapid fermentation and gas production. For this reason, industrial researchers are also working to perfect the art of feed particle size!

Antibiotic-pesticides may alter the stomach's bacterial populations and its ability to relieve trapped gas. In addition to routine antibiotic use to boost productivity and efficiency, animals are consistently being dosed with low levels of pesticides that have antibiotic properties. This is because our regulatory and research institutions have largely overlooked the fact that several widely marketed pesticides also act as "biocides." These antibiotic chemicals are then applied to millions of acres of plants destined for food and animal consumption.

My father becomes rather irate when scientists blame animal agriculture for growing incidences of antibiotic resistance. While the media berates livestock farmers' routine use of antibiotic drugs, the routine use of pesticide-antibiotics to treat plants goes largely unmentioned.

How exactly does a pesticide-antibiotic harm an animal's health? Let's use glyphosate as an example. In addition to the herbicide and chelation patents mentioned previously, there is a third, less widely recognized patent on glyphosate. This is patent #7771736—a patent permitting its use in the medical treatment of pathogenic organisms including parasites.[3] For the same reason that glyphosate works as a "broad-spectrum" herbicide, glyphosate works as a "broad-spectrum biocide." It reduces some bacteria and parasitic organisms' abilities to carry out important metabolic processes. There are too many "susceptible organisms" listed in the patent to include them all here. Each susceptible family of bacteria and parasites sits adjacent to a disclaimer that glyphosate's effects are "not limited to" the organisms on the list. The patent author essentially admits that he or she does not know the full range of effects of the company's own patent! Of course, there are many bacteria we know very little about. In the cow's gut alone, there are hundreds of species that a broad-spectrum antibiotic-pesticide could potentially upset.

Interestingly, several of the bacteria affected by glyphosate may be causing an increasing number of intestinal malabsorption problems in both human and domestic animal populations. To date, several types of highly pathogenic bacteria are known to be resistant to glyphosate. These harmful bacteria include several varieties of salmonella and clostridia. Clostridia, for example, can overpopulate the intestines in the presence of glyphosate and are increasingly recognized as problems in both large animal medicine and human medicine.

On a similar token, numerous bacteria that are beneficial to a cow's digestive processes are negatively affected by glyphosate. In the presence of glyphosate, research has shown that the numbers of enterococcus faecalis, enterococcus faecium, bacillus badius, bifidobacterium adolescentis, and lactobacillus all decline.[4-6] Each of these good bacteria help the cow to digest its food and help reduce the risks imposed by other bacterial residents. When a cow eats feed products containing glyphosate residues, these residues may lend to increased incidences of illness while negatively impacting normal digestion and absorption.

Because glyphosate is toxic to several beneficial enterococcus species that work to quell the negative impacts of normal clostridial populations, this antibiotic-herbicide could be a significant predisposing factor for several clostridial diseases.[6] Clostridial diseases can wreak havoc on one's guts, causing bloody diarrhea, intestinal hemorrhaging, fever, weight loss, and, occasionally, death. Certain types of clostridia can also produce toxins that affect one's brain and muscle coordination. Because clostridia are a common bacterium found in many environments, they can also be hard to treat and control.

In my father's practice both Clostridia difficile and Clostridia perfringens are becoming more common. One strain of C. perfringens can cause severe forms of bloody bowel syndrome. When C. perfringens Type C affects piglets, they can have sudden and often massive intestinal hemorrhages followed by collapse and death at just one to three days old. C. perfringens type A, on the other hand, can cause intestinal hemorrhaging and chronic diarrhea but is usually less severe.

C. difficile is particularly problematic because most antibiotics don't kill it. In fact, many doctors suspect that increasing C. difficile populations are brought on by the overuse of antibiotics. Broad-spectrum antibiotics may wipe out the healthy enterococcus species as well as target pathogens, canceling the body's system of checks and balances. In animal medicine, C. difficile may be the most common cause of New Neonatal Porcine Diarrhea (NNPD). Yep. That's right! In addition to Post Weaning Failure to Thrive Syndrome (PFTS), baby pigs are also struggling with chronic diarrhea, two ridiculous diseases that may be more connected than most researchers think.

According to Denmark's National Veterinary Institute, pigs with NNPD contract diarrhea within the first week of life. They then lose weight and have a difficult time catching up with their peers. Unlike other forms of piglet diarrhea, antibiotic and vaccine interventions appear to have little effect on NNPD. Researchers are hoping to identify the source of bacterial imbalances to reduce the consequences of this spreading disease.[5]

Unfortunately, some groups of veterinary and food safety researchers report growing concerns that C. difficile infections could be acquired via interspecies transmissions between animals and humans. Apparently, the incidence and severity of C. difficile infections have increased in both human and animal populations and there is a large overlap between the strains involved.[6]

After analyzing C. difficile research, one German research team proposed that additional research is necessary to ascertain what is causing increased incidences and severity of C. difficile infections. They also suggested that "competitive exclusion" of toxic strains appears to be the most beneficial way of treating humans and animals. In fact, they reported that one of the most important steps in treating C. difficile may be to withhold antibiotics because antibiotics disrupt healthy balances in the gut.[6] They cited several trials where animals benefited from a healthy amount of competition between bacteria.

Interestingly, a few of my fathers clients tell him that the only way they've been able to manage their pigs' clostridial bloody bowel infections is by regularly administering a bleach solution to infected animals! Many of the antibiotics they historically relied on no longer work. My father points out that the only reason this is effective is because livestock animals already have shortened lifespans. Treating a creature with regular doses of bleach wouldn't ever be practical or useful otherwise.

Bleach is also used in human hospitals where workers are instructed to wipe down surfaces with bleach solutions and to wash their hands to prevent the spread of pathogens. Still, rates of C. difficile have increased dramatically in both adults and children. According to one Mayo Clinic study, the rates of C. difficile infections in children rose from 2.6 cases per every 100,000 to 32.6 cases per 100,000 between 1991 and 2009. It seems important to note that approximately seventy-five percent of these cases developed outside the hospital.[5]

My father is concerned that antibiotic-pesticides may be playing a larger role in human and animal illnesses than many health professionals think. When feeds are contaminated with adequate concentrations of antibiotic-pesticides, both cows and humans, eating poorly balanced diets, face another health challenge. These edible antibiotics further disrupt their guts.

This problem could occur when farmers use antibiotic pesticides like glyphosate as drying agents on feed crops. When residues on treated corn and cereal products are ingested, they do not need to be metabolized to start trouble; they simply work on susceptible bugs in the gut. Public health researchers, agronomists, and soil specialists should be working together to better evaluate antibiotic-pesticides and their effects on antibiotic resistance in soils, plants, animals, and humans.

The same German team working to highlight the importance of

competitive exclusion in C. difficile cases is investigating an even deadlier clostridial disease called chronic botulism. Chronic botulism can cause cows to experience constipation, diarrhea, lameness, paralysis, edema, and death. C. botuli produce toxic spores that quickly work to degrade a cow's nervous system. These spores also leave a cow's stomach lining irritated and inflamed.

In fact, Clostridium botulinum produces some of the most potent toxins known to man. These toxins can affect various animal species. Seven members of the C. botulinum family have been recognized and are named A, B, C (1 &2), D, E, F, and G. Types A, B, and F can cause human botulism while types C and D cause most cases of botulism in domestic animals. Some of the toxins released are so damaging that they have been investigated for use in biological warfare.[4,6]

When glyphosate puts stress on a cow's microbiome, the concentration of potentially harmful C. botulinum populations soar. These gut residents produce far more bont neurotoxin than would normally be present in a cow's stomach. In fact, one study found that with just .01 mg/ml of glyphosate ingested, the amount of bont neurotoxin in cattle rose from less than 1 ng/ml to more than 3000 ng/ml of neurotoxin present![4,5,7] When cows absorb enough of this toxin, they become partially paralyzed and die.

While past data suggests that separate strains of C. botulinum are responsible for human and animal cases, researchers are reevaluating the conditions that have lead to cases affecting farmers in close contact with cattle. Recent research indicates that a new form of chronic visceral botulism may affect both humans and animals. This research was conducted after several German cattle farmers developed botulism themselves.

The farmers who contracted botulism reported dizziness, blurred vision, fatigue, speech problems, difficulty breathing and swallowing, and overall weakness. After analyzing the bacterial strains involved, researchers were unable to discern if transmission from cattle to humans had occurred. However, several researchers still suspect that cattle, who generally harbor numerous strains of C. botuli, could have passed some form of botulism on to these farmers.[7,8]

Yet, many of our domestic animals are eating antibiotic-pesticide residues well-above established toxicity thresholds on a regular basis! This practice is likely wiping out the competition needed to keep both disease and resistant bacteria at bay! A need for healthy competition to prevent clostridial diseases seems especially important when diseases like infant botulinum and c. difficile occur after such creatures take over one's digestive tract.

This information also means that we need to be more aware of the antibiotic resistant genes and bugs accumulating in our soils. Because they can reproduce rapidly, bacteria can resist undesirable conditions more easily than humans. Those who can survive a toxin in the environment will pass their genes down the line. For some species of bacteria, weeding out the survivors from the failures takes minutes. For humans, this process takes decades, and may still be problematic even with advanced technologies.

Only the strongest, most resistant bacteria survive antibiotics designed to kill them. Unfortunately, when humans make new, more powerful antibiotics we have to be prepared to deal with a more powerful resistance. In addition to various antibiotic resistances observed in the U.S., scientists at Newcastle University in northeastern England have reported a rise of antibiotic resistant genes in our soils. Some of these genes are associated with human health problems. Researchers found that 78% of resistant genes from four antibiotic groups showed increased resistance levels since 1940.[9,10]Is there any chance that these resistance levels may be linked to the millions of pounds of antibiotic-pesticides we've been spraying across these soils for years?

While we can visibly see glyphosate resistant weeds, we have no easy, inexpensive way to document antibiotic-pesticides' overall effects on soil and plant health. Yet, we can study the side effects. In their paper describing interrelationships between glyphosate and soil health, Johal and Huber document more than thirty plant diseases that have increased in conservation agriculture systems due to soil and plant pathogen disruption![11,12]

Clostridia don't just compete with the thousands of bacteria lining a cow's gut, they also live in our soils and compete with other healthy and harmful bugs underground. When these microbiomes, sitting at the base of all food chains, are disturbed, we begin to feel the gaseous, sometimes toxic, and deadly consequences. Only when we begin to better appreciate how interconnected our health is to the health of the plants, animals, and land around us, can we make true agricultural progress. Perhaps we should start measuring industrial ag efficiencies in pounds of balanced bacterial populations in a healthy human gut?

I have now successfully discussed germs, disease, diarrhea, farting, and belching all in a single chapter. My middle schoolers would be so proud! If there weren't so much farting and belching going on, I'd tell you to take a deep breath before we move forward. As usual, this is a lot to digest!

DIAGNOSIS 7: LISTERIOSIS

Being the daughter of a large animal veterinarian isn't always fun and exciting. In fact, I sometimes resented my father for his obsession with his work. His work never ended. For perhaps an unusually long time, I had the impression that one's profession is every human being's primary focus: when work calls, all other priorities aside! It's time to fulfill your call of duty!

Though things have slowed down over the years, my father still takes vet calls at all hours of the night, evening, and weekend. Today he swaps every other weekend and vacation days with his partner. Still, someone must always be on "duty" or ready to fill in on busy days. When I was little, my dad got every third weekend off and typically took a one or two-week vacation in the summer.

Despite the time he spent away from home, I got a reputation for being a "daddy's girl." I followed him around outside when he was home and, like my older brother, went on vet calls with him when I was very young. Farm wives would take me in and feed me cookies while Dad talked somebody's ear off about this or that ailment. Sometimes, clients' kids would show me the new puppy or their favorite barn cat when I got bored.

Of course, part of the tagging along bit was because I wanted to spend time with my Dad. But I also wanted to understand what was so important about his work that it demanded so much time away. After just a few veterinary experiences, I stopped asking this ridiculous question. His absence made perfect sense. While an 80 hour work-week is not my typical recommendation for folks, a veterinarian doesn't have the liberty to call in sick or maintain scheduled hours when an animal is dying just ten miles down the road.

For these reasons, I was a bit anxious about my father's workload the week of my wedding. While my father had taken my wedding day off, he was scheduled to work the rest of the week. I wanted to spend time with him but knew that I'd have to make myself useful to do so. With preparations and all, I wasn't quite sure I wanted to commit to a never-ending day of vet calls just to spend time with Dad. Then again, not spending time with him before my big day didn't feel right either. It was this line of thinking that almost had me killed just two days prior to my wedding.

I made a compromise. I would help treat a D.A., then Dad could drop me off at home before heading to his next vet call. We could catch up on the way there and possibly grab lunch at a diner in Coggon. Inevitably, another important vet call popped up just down the road. My commitment to one call turned into two and then three. On the way to the fourth, I expressed growing frustrations that I had *so much to do for the wedding and could you please just run me home after this call even though it's out of the way FOR THE LOVE OF GOD?*

"Yes," he promised. That could be done.

Much to my chagrin, the farmer managing the fourth farm informed us that he had two cows to look at rather than one. The first cow was "bad off," and "there probably wasn't much left to do." We walked around a dairy to find a cow lying in the grass between a holding pen and a cement wall. She was lying on her side, panting heavily, and salivating like crazy. Once in a while, she'd loll her head to the side, then let it roll back down to the ground. The skin folds around her face sagged and her lower jaw drooped open.

My father quickly decided that she had listeriosis and informed the farmer that he was probably right. She was going to die. The antibiotics they had administered the day before were too late. Best to make her as comfortable as possible and wait for the inevitable.

I spent the next few minutes daydreaming about my wedding while Dad chatted with the farmer about listeriosis. The next thing I knew, the 1200 pound creature had risen to all fours and started charging me toward the cement wall!

I have had many close encounters with serious injury over the years. Nothing super special; I think most rural kids take a few serious knocks here and there. I have fallen off a donkey onto highway pavement and I've curled up with my sister under a calf's body on a -20 degree day to stay warm. I've also been flung off a horse mid-surgery before taking an unexpected ride around a client's house on this same animal, clinging to its mane for dear life. While each of these incidents had the potential to end poorly, none rattled me as much as this listeriotic cow.

When I saw the cow get up, I was rather confused. It circled once, then twice, then seemed to hone in on me before lowering its head and charging at my legs. If I hadn't caught the edge of a large round bale out of the corner of my eye, I would have been a dead woman. Luckily, the energy directed toward my legs dissipated when the cow suddenly became distracted and began circling: spinning round and round like a wound up top. I reached the bale and felt some intense pressure against my right thigh as it rammed into me the first time. Then I attempted to squirm up the bale. By the time the cow was done with its next circle, my left leg was high enough that I could push off its lowered head when it came at me again. With the momentum from one swift thrust, I propelled myself to the top of the bale, grasping at twine and straw all the way up.

My father and the client watched in disbelief as the cow continued to ram into the bale and circle for several minutes. Meanwhile, I struggled to regain my composure. Then my Dad, who is a bonafide genius when it comes to veterinary medicine, but sometimes the least tactful human being on earth, shook his head and laughed, "Thank God that wasn't me!" he said.

While he had meant that he wouldn't have been quick enough to scramble up the bale, I wasn't in the mood to appreciate anything other than concern. "Thanks a lot!" I scowled back. Then I settled into silent anger, struggling to choke back frightened sobs. The client and his farmhand eventually secured the cow.

I felt some sense of justice and slapstick relief a few minutes later when my father accidentally walked directly into a metal gate while preparing to examine the second cow. We made it home half an hour later: both bruised and rather irritated. Talk about a lovely pre-wedding father-daughter date!

Neither of us felt as bad as the cow. The cow was busy dying from listeria, a bacterial infection that can affect a wide variety of animals including humans.

When listeria are ingested or inhaled, they can cause infection, early-term abortion, and septacemia (blood poisoning). When listeria penetrate the tissues, as they did with this cow, they can damage the intestinal walls, brain-stem, and placenta.

Swelling and irritation of the brain, called encephalitis, is the most widely recognized form of listeriosis in ruminants. If a listeriotic cow is treated with aggressive antibiotics within the first day or two, the cow might survive. If not, and the damage is great enough, the cow will die. The current recovery rate amongst encephalatic cows is about 50%.[1]

The symptoms of encephalitic listeria are generally onset by bacterial lesions that damage the cranial nerves and brain. A cow will become disoriented and begin to circle as we saw that day. It may also lose control of its facial muscles, and less frequently, experience involuntary running movements, like the charging incident that almost sent me to my grave.

What I find most fascinating about listeriosis is that the same bacterial strain that causes these visibly crazy symptoms can be isolated from the intestines of many healthy populations. Listeriosis regularly populates plants, soil, water, milk, cheese, birds, fish, humans, and other intestinal mammals.

Because we know that it can be transmitted via feces and infected tissues, doctors advise certain individuals to steer clear of animals exhibiting harmful symptoms even though they may already have some amount of listeria in their guts. Specifically, people with compromised immune conditions are encouraged to avoid contact with listeria. Even though they are rare, there have been documented incidences of animal to human transmissions.[1]

It appears that some threshold number of bacteria must populate our guts before a systemic outbreak occurs. So what upsets this balance exactly? When is listeria okay and when does it get out of hand?

Historically, these questions have been important to American dairy farmers. Prior to distribution, most American dairy products, including milk, are pasteurized. During pasteurization these dairy products are heat-treated to prevent bacterial growth and to extend a product's shelf life. Pasteurization became common in the late 1800s and early 1900s as cities became more densely populated and fewer people had access to fresh dairy products. Because, numerous bacteria can grow on milk, including several potentially deadly varieties, like listeria, E. coli, and salmonella, the sale of unpastuerized, "raw" milk is prohibited and discouraged in many areas.[2]

In the United States, most states impose restrictions on raw milk or ban the sale of raw milk altogether. While our regulatory agencies argue that these restrictions help prevent illness, raw milk proponents argue that pasteurization damages nutrients and beneficial organisms in the milk.[3-6] Some independently owned and operated dairies also dislike the fact that they can no longer sell milk directly to consumers. Because these farmers have to send their milk to manufacturers for pasteurization, there is a need for a middle-man which eats into profits.

Recently, my husband's high-school friend, James, decided he would join us for our annual summer trip to Iowa. He grew up in a suburb of Boston and was excited to see a piece of real farm country. Naturally, my family was excited to have another city boy to pick on and tried its best to educate him in all rural matters. We sent him on several vet calls, walked him through a couple of small towns, subjected him to my father's veterinary diatribes, and fed him up family style. By the end of the trip, he was equally skilled at ribbing my family about our lack of knowledge regarding urban matters.

On the last evening of his visit, James wondered if it would be too much trouble to go out for some local ice cream. He had been to several dairies throughout his stay and wanted to enjoy some local flavor. My parents, who are very proud of their farm backgrounds, couldn't think of a single place that served locally made ice cream within reasonable driving distance. We could drive to the nearest city, about an hour away, and see if we could find some there?

I was surprised by the feeling I associated with this question. It was one of slight embarrassment and I wondered if my parents felt it too. Here we were, acting like know-it-alls to show off to an urbanite, and we couldn't think of a single place serving the products of our pride. Beyond feeling somewhat embarrassed, I felt irritated. Who are rural farmers working for when they can't drink the milk produced on their own farms? Or when no one on these farms even cares to taste or sell the end-product of their labors?

Instead of going out for local ice cream, we ate some pasteurized, frozen Blue Bunny ice cream and Mom told stories about how her family used to drink milk produced on their dairy. The whole situation struck me as strange and I began to ask lots of questions. During the conversation, I learned that my sister-in-law, who grew up on one of the only certified organic dairies in the county, didn't grow up drinking her family's own milk products! Even though they were organic, these products had to be pasteurized prior to sale. Some family members felt that by drinking their own milk they would be taking a financial loss. They recognized that they could sell a gallon of organic for more than it would cost to buy a gallon of conventionally produced milk at the store. I was surprised. I know that our ag systems are messed up, but such practices seem especially strange. What would our grandparents think? They grew up drinking and eating much of what was grown on Midwestern farms.

While my dad is pretty well convinced that pasteurization is necessary to control harmful bacteria, especially when bacteria-heavy silage is being fed to modern dairy cows, I can't make up my mind. Like other two-headed issues, I can see strong cases for and against pasteurization. The research describing raw milk's health risks is wrought with inconsistencies. My older brother has decided to serve his family raw milk products despite the FDA and USDA's warnings against doing so, as well as concerns expressed by his veterinarian father.

My brother is not alone. A number of Americans are returning to the practice of buying raw milk products directly off American farms. They point to research that indicates that raw milk, coming from well-managed, grass-fed farms, is hardly as dangerous as American regulatory agencies and dairy conglomerates want us to believe.

According to Mark McAfee, of Organic Pastures Dairy, California state inspectors have never found salmonella, listeria, or E. coli (0157:H7) in Organic Pasture's raw milk products. Tests on Organic Pastures milk also revealed that when these same pathogens were added to raw milk samples at low levels, pathogen numbers declined rather than grew! It appears that organic, raw milk, coming from healthy animals, may actually reduce some pathogen numbers.[3] Additional research demonstrates that grass-fed cows, receiving relatively low levels of antibiotics, are less likely to slough off pathogens in both their manure and milk. This indicates that our listeriosis concerns are likely heightened by the ways we feed and house conventional dairy cows.[3-6]

These findings are actually consistent with veterinary descriptions of listeria infections. In the animal community, listeriosis is primarily associated with winter and spring outbreaks amongst ruminants housed in tight buildings and feedlots. According to my father, spoiled silage that has sat around in piles all winter is usually ripe with bacterial growth. Removing or changing the silage in a cow's diet often stops the spread of listeriosis altogether while feeding the same silage months after the infection can cause new cases. It appears that cattle, like humans, like to eat fresh greens rather than highly processed, rotten greens. Once again, poor feed quality is a factor in bacterial infections.

While we have become highly attuned to the risks presented by certain pathogens, we have ignored the benefits consumers once obtained while drinking raw milk. This milk contains several enzymes that kill pathogens. In fact, the Food and Drug administration has approved the use of lactoferrin, one naturally occurring milk enzyme, to kill bacteria in beef slaughter plants.

Experts who remind new parents that breast milk is as much a medicine as a food could say the same thing about cow's milk. According to McAfee, "It is no wonder that dairy plants that pasteurize must be kept absolutely spotless. There are no remaining safety systems in the processed milk."[3] Both the health-promoting bacteria and pathogenic bacteria die when the milk is treated.

Because healthy bacteria are important factors in warding off disease, I wonder what the health costs are to calves weaned so early into a world ripe with possible pathogens. Do milk replacer formulas, fed to baby calves who are later cooped up in high-density living spaces, pass on any of the beneficial enzymes and bugs needed to boost a calf's immune system? Perhaps raw milk substitutes only exacerbate our antibiotic dependencies because we have eliminated the body's natural antibiotics? In these cases, nature likely knows much more than we give her credit for.

According to food safety progress reports, the incidences of foodborne illness have declined somewhat in the last one hundred years and modern pasteurization methods kill 99.99% of bacteria.[7] Unfortunately, these reports include little data on how illnesses related to the reduction of health-promoting bacteria and enzyme populations have changed during this same time. I find it difficult to trust many food safety studies given my knowledge of what is tested and what isn't. Industrial ag proponents argue that our food safety systems are working wonders while ignoring the incredibly complex multi-species immune systems that keep humans and animals healthy.

It seems odd to most Americans then that countries like France have been selling and consuming raw milk for hundreds of years because, in general, we are so adamant that pasteurization is necessary. Many foreigners, on the other hand, consider raw milk and cheese to be of the highest standard. Cheese makers from around the world have taken advantage of milk's beneficial bugs to make interesting, value-added dairy products. Some of America's highly processed and pasteurized cheeses seem bland in comparison.

In my father's opinion we may not be able to reduce bacterial infections, like listeriosis, without compromising herd health, because America no longer relies on regionalized food systems. In regionalized food systems bacterial balances can be more easily observed. While I would feel comfortable drinking raw milk from several Delaware County bulk tanks where the animals live in low-stress environments and are fed well, I cannot say the same for others. Unfortunately, many of Delaware County's uniquely managed farms churn out milk and butter that only ends up on the same shelf, under the same name. The new goal to mass market as many cheap products as possible, for the sake of global distribution, is swallowing both our tastes and our pride.

Today, our abilities to trust the safety of our food sources is tangled up in a second, two-headed debate: one between those who think buying local will save the planet and those who advocate that globalized systems of mass production must save us first.

The day after I was almost killed by the listeriotic cow, I read an article in the *Hoard's Dairyman* entitled "Can Buying Local Food Really Save the Planet?" The article was written by Jude Capper, an assistant professor at Washington State University. In the article, Capper wonders why consumers increasingly assume that eating local can be anymore environmentally friendly then eating conventionally. She discusses the misguided assumption that local foods require fewer inputs while forwarding that this view only furthers foolish sustainability *trends*.[8]

I found myself thinking about my family's inability to find any locally produced ice-cream in an area that used to be heavily populated with independent, raw-milk dairies. Is their any truth to Capper's remarks?

To prove her point, Capper compares three types of food purchases: buying direct from a farm, using a buyer's club, and shopping at a local grocery store. She argues that those participating in the local food movement often overlook more productive trips to and from large retailers when making their food purchases. I would agree that this could be a problem, especially if a consumer makes multiple, rather lengthy trips to numerous farms.

In her analysis, Capper compares a CAFO (a Confined Animal Feeding Operation) raising hundreds of thousands of chickens to a farm direct example. She explains that the CAFO can transport 23,400 dozen eggs per tractor-trailer load while traveling about 2,405 total *food miles*. When she does the math, this works out to about .24 gallons of fuel per dozen eggs. She later compares this to a farm-direct example, in which one person travels 300 total miles to pickup one dozen eggs! This works out to be about 13.2 gals/dozen, a 55-fold increase in gas-consumption over the CAFO model![8]

While Capper admits that her example "is a simplistic rendering," her article leaves one with the impression that she is conducting a sound scientific investigation as opposed to one of those crazy non-scientific studies touted by local food supporters.[8] It also makes one wonder why any idiot would ever choose to shop locally provided the sheer inefficiencies involved in buying a local product.

In conclusion, Capper states that,

> Demonizing highly efficient production and transportation systems through marketing efforts plays into popular misconceptions. It is imperative that consumers have the freedom to enjoy a selection of production systems and buying choices. Consumers making purchases based on environmental criteria should have all the facts. This includes sound science, rather than simple philosophical assumptions that ultimately lead to greater resource use and carbon emissions (pg 43).[8]

She is dead on. People should base their food choices on sound science and scientifically valid statements. Unfortunately, I would not include her own analysis or assumptions amongst these criteria.

To begin with, Capper reduces one's desire to purchase locally produced goods to a *trend* and a matter of gas mileage. Comparison of food consumption patterns is based solely on fossil fuels used for transport to and from a supermarket. Perhaps Capper is unaware of the other reasons one might choose to support a local farmer or business over larger corporations? Perhaps she is also unaware of other on-farm practices that might require fossil fuel inputs?

While Capper brings up several interesting questions, she, like other industrial ag proponents, ignores many common sense inputs and outputs involved in modern food production. She also makes several assumptions about local food supporters and environmentalists.

I cannot count the number of efficiency studies that ignore the fuels required to fertilize conventional feed, spray it with pesticides, manufacture pesticides, produce nutrient supplements, deliver and transport feed, process meat, dispose of animal wastes, and package food products. Many of the industrial life-cycle analyses designed to ensure you and I that our food products are produced more efficiently and sustainably have left out these key data points altogether.

Such reports often underestimate those who have willingly spent time thinking about their food purchases as well as individuals who have chosen quality over personal economic gain. Not only have many consumers considered a more full-range of fossil fuel impacts, they also have values, beyond environmental reasons, that influence their food choices. Some of these people shop locally because they prefer to create demand for a food product that is fresh, relatively energy efficient, mindful of animal welfare, and, overall, a more environmentally friendly food. Some of these individuals walk or bike to farm pick-ups to buy their local food products. Few Americans have the time or are dumb enough to drive 300 miles just to purchase one dozen eggs! This would never make economic sense and would conflict with the other values many local food eaters hope to support!

While science can be an important tool, it is just that: a tool. We need to draw on our common sense, values, and personal experiences to make important decisions while weeding out the really bad science from the decent to good.

As I have mentioned before, no two farms are alike. I have seen large conventional dairies managed in a way that is much more in line with my values than small, poorly operated dairies. By and large, however, this is not the case. Nonetheless, I have to admit that my observations are not necessarily consistent with every scientific generalization. Then again, one needn't drive 300 miles to support local farms in ways that are much less harmful than Capper implies.

Supporting a local farm is not unlike supporting any other local business. Just as the previous analysis failed to address many farm inputs, it also failed to address important outputs. How much of what each farm produces in economic profit remains in the community to provide local wealth and jobs? How much of what's produced on a locally owned dairy can generate business if the milk produced there becomes tasty locally owned and sold ice cream? Industry funded studies that ignore connections between local farms and local economies are regionally and fiscally near-sighted.

Those areas with the most local farms tend to have resilient economies. These areas also employ more local citizens, while providing citizens with a healthier array of fresh food options. When farms are well-managed and diverse, humans and the environment can benefit as well.

Iowa has a reputation for having some of the most fertile soils in the nation, prime growing land for many fruits and vegetables. Yet, what we think of as the "Food Capital of the World" can't even feed itself. Eighty-five percent of the nine billion dollars Iowans spend on food leaves the state.[9,10] This need for food imports is a direct result of farm policies that encourage high amounts of corn and bean production, most of which are geared toward producing ethanol and animal feed.[9]

What is produced provides questionable health, a matter frequently observed by doctors and nutritionists as our corn-fed waistlines go the way of our super-sized beverages and meals. In fact, most of the 150 pounds of sugar the average American consumes each year, a noticeable hike up from the 4 pounds we consumed in 1750, is corn based.[11] Feeding Iowa on corn alone isn't doing anyone a service. I find it somewhat insulting even. Given the amount of agricultural land available, the possibilities to diversify local diets while improving our livelihoods are tremendous!

U.S. Corn Use - 2011[12]

Usage Type	**Percent of Total**
Ethanol	39.5
Feed and Residual	36.4
Export	13.0
High Fructose Corn Syrup	4.1
Sweeteners	2.1
Starch	2.1
Beverage Alcohol	1.1
Cereals/Other	1.6
Seed	.2

These possibilities hold true for other states as well. According to USDA data, if every American followed USDA dietary guidelines, the U.S. would need to produce 7.6 million more acres of fruit, 6.5 million more acres of vegetables, and 11 billion more pounds of milk to satisfy its needs.[10.] Yet today we rely on imports for many of our fruit and vegetables while providing federal support for crops we already produce in abundance!

Small political changes could have huge impacts in this system. Americans could easily produce enough fruits and vegetables to feed American citizens. In fact, research demonstrates that relatively little land would have to come out of crop production to produce enough fruits and vegetables to feed both Iowans and surrounding urban areas. One study concluded that for every job displaced in corn and bean production, five new jobs could be created to run fruit and vegetable farms. As it stands, our fruit and vegetable imports are on the rise, more than doubling in the last ten years. Meanwhile, only 0.05% of Iowa's arable land is planted to fruits and vegetables.[2] We continue to rely on the same land use patterns despite the health and economic costs to our nation's states and communities.

According to a study conducted by Ken Meter of the Crossroads Research Center, the net income of commodity crop production in many rural areas is actually negative! Francis Thicke explains how this applies to Iowan economics in his book A New Vision For Iowa Food and Agriculture:

> In an eight county region of northeast Iowa, farmers sold an average of 1.08 billion worth of crop and livestock commodities per year during 1999 through 2003. Yet they spent $1.14 billion to raise those commodities. The result was an average loss of $62 million in production costs for the region each year. That loss was offset by an average of $173 million of federal subsidies, and $72 million of other farm-related income (primarily custom work and rental income) each year. During the same timeframe, consumers in the eight county region spent $400 million per year buying food imported from outside

the region. (pg 147-148).[10]

This amounted to, what Thicke, called a *large loss in potential economic growth.*[10]

Iowa's fruit and vegetable farmers work against this grain. This choice is often financially difficult given current regulatory frameworks. Historically, crop insurance has not been an appealing choice for many fruit and vegetable farmers because the threshold for loss, which has largely been based on acreage, is too high and the payout is too low for those managing so few acres.

My brother, who has several acres in vegetable production, said it hasn't really been worth it for him to pay for federally supported insurances. In the past, insurers have used corn and bean baselines to estimate the potential value of his crop. Two years ago he was told that his farm would be insurable for up to $12,000, even though his fruit and vegetable crop was worth several times this amount—not to mention the 80 or so hours worth of work he did per week. Such policies only add insult to injury. For the diverse, small acreage farmer, investments in seed and infrastructure seem like better insurances than subsidized policies that may or may not pay out.

Some states are working to address these disparities. The state of Illinois, for example, passed the Illinois Local Food, Farms, and Jobs Act. By 2020, 20% of all food products purchased by Illinois state prisons, universities and agencies will be produced in Illinois. Ten percent of all food products supported by state-dollars will also be produced within the state.[13]At the state level, the income generated should result in net benefits for all citizens. At the national level, such measures could decrease our dependencies on the national government and foreign food suppliers. If you think about it, it's kind of disturbing that we have to create an act just to reach goals of 10% and 20% in major agricultural regions!

A similar bill was proposed as part of the last Farm Bill legislation.

Representative Pingree of Maine and Senator Brown of Ohio introduced a National Local Farms, Food and Jobs Act to improve economic opportunities for farmers and increase access to fresh, local foods.[13]The bill is full of measures that could work at the state level, as well as measures to help develop national insurance policies that support on-farm diversity. These intelligent policies will increase food security at home while reducing other costly items, such as health-care, national defense, and emergency aide.

I have been fortunate to live in several areas where local food systems thrive and excuses for two-headedness don't hold up. During the recent economic downturn, the number of local farms and businesses actually increased rather than decreased throughout the Pioneer Valley. A local farm neighbor, Mountain View Farm, now grows enough food on 60-70 acres to provide for 1400 share-holding families in the summer and fall as well as 400 additional share-holders in the winter. The workers here grow between 500,000-750,000 pounds of fresh fruit and vegetables each year for these community members. None of this yield is being used to feed cars. Some, approximately 100,000 lbs, *is* donated to the area food bank.[14] It is farms like these that provide a diverse supply of human and animal food, that create economic security.

Likewise, communities with strong local banks and locally owned businesses are typically less vulnerable to price changes and high unemployment rates than areas without strong regional supports. This is as visible in Decorah, Iowa, where I attended college and worked with various local farms, as it is in the Pioneer Valley. Diversity and healthy competition between businesses and bacteria visibly create resilience.

Yet those Americans taking time to put nutrition on their priority lists are still disparaged by industrial ag advocates. These people treat health concerns like *trends,* reducing changes that take personal commitment to the next latest thing. Some of the national animal diseases we hear about in the news, including listeria, wouldn't be nearly as alarming if we could manage them better at regional levels. Today, we discuss these problems in globalized contexts and refer to the solutions we've used to fix them as trends—as though someone's decision to feed a family quality meals or to support a local business is comparable to someone's decision to rock the latest fashion! In so doing, we fail to recognize the time and effort many Americans are making to keep their families, environments, and communities healthy.

DIAGNOSIS 8: MYCOTOXICOSIS

In addition to the important work bacteria do in our digestive systems, bacteria help detoxify mycotoxins in our food. Mycotoxins are produced by molds that contaminate feed crops during growth, harvest, processing, and storage. Even at low levels mycotoxins can increase disease susceptibility and reduce performance. At high levels mycotoxins are linked to a wide range of conditions including allergic reactions, poor digestion, reproductive disorders, organ damage, and, in worst-case scenarios, death.

A single fungus can produce several different mycotoxins under various conditions. Fungi interacting with other fungi in multiple feed ingredients can produce even more toxins. An animal eating mycotoxins in large quantities will need a strong immune system to stay healthy, a function largely dependent on bugs in the gut and liver.

Typically, an animal's immune system has to be fairly compromised for fungal toxins to cause observable damage. A family friend recently contracted a serious fungal infection and we became worried that he might have cancer. Cancer victims tend to be most susceptible to fungal diseases because their immune systems are already fighting cancer causing agents and the chemicals used to treat them.

So, when my father began to see mycotoxin induced diseases in greater than normal quantities, he knew he was either dealing with very high mycotoxin levels or compromised immune systems. In the world of veterinary medicine, mycotoxin induced diseases are referred to as mycotoxicosis.

While there are many toxic molds that produce mycotoxins, only a few mold families, particularly those affecting grains like corn and wheat, significantly affect animals. One of the most widely researched pathogenic mold families is fusarium. Several of the previously mentioned plant diseases are actually caused by fusarium fungi. These molds are present to some degree in most major cash crops.

The fusarium mold family produces four types of toxins: zearalenone, vomitoxin, fumonison, and T2. Each is associated with different symptoms when mycotoxicosis occurs. For example, in pigs zearalenone acts as an estrogen mimic, disrupting hormonal processes while vomitoxin affects digestion and can drop feed intake. Fumonison, on the other hand, can damage hog hearts and cause pulmonary edema, while T2 ruins the liver, occasionally destroying the pig's immune system. As was the case with clostridia, some mycotoxins have been researched for potential use in warfare. In fact, in the 1970s, the Soviets harnessed the damaging effects of the T2 mycotoxin for use as a biological weapon.[1]

Today my dad is treating more herds with mycotoxicosis. In one herd, fumonison levels were so bad, pigs died of heart complications. This was evident in postmortem analysis and lab tests. When evaluating potential cases of mycotoxicosis, my father typically sends feed samples into the lab for confirmation. Over the years he has made the following observations:

1. In tested herds, the four toxins produced by fusariums have often been "marginally safe" or "potentially unsafe." This was not the case a decade ago.

2. Clients feeding their animals glyphosate treated crops have had higher levels of fusarium mycotoxins in their home grown

feedstuffs than clients using *non*-GM, *non*-glyphosate treated feeds (including feeds high in plant matter like hay and alfalfa). This has been especially true during cool, wet years.

3. Animals fed fermented, ethanol byproducts have regularly had up to three times the levels of fusarium mycotoxins in their feed products when compared to the original corn grains used to make this ethanol.

4. By reducing distiller grains in the diet, replacing glyphosate treated feeds, and administering enzymes to transform mycotoxins, most herds can fully recover.

For these reasons, my father began to suspect that fusarium molds grow in larger quantities on glyphosate treated, GM crops than on other crops. His research has steered him down a familiar path. But the path keeps going. He has found people from halfway around the world with similar concerns about the ways pathogenic fungi affect feed crops.

Fungi, like bacteria, are found in soils and form symbiotic associations with plants. The benefits of these relations include improved nutrient uptake, soil quality, nitrogen fixation, and resilience. A healthy fungal population, like a healthy bacterial population, contributes to the health of all terrestrial plants and animals. And like bacteria, harmful populations can be managed by healthy counterparts.

As fungal outbreaks become more common, more scientists are becoming interested in fungi. In fact, many of the recently reviewed, disease-driven extinctions can be accredited to fungi.[2] Fungal diseases have negatively affected crops, plants, bats, frogs, humans, and snails.

One type of fungus increasing throughout the world is called Bd (*Batrachochytrium dendrobatidis)*. This fungus attacks amphibians via the skin, disrupting water intake to such an extent that amphibians can die of heart failure. Unfortunately, Bd has been found in hundreds of species of amphibians around the world. Some Central American areas have lost almost half of their amphibian populations.[2]

In the United States, land managers have closed caves to protect remaining bat populations from another deadly fungus called G. destructans. This fungus is thought to cause White-nose syndrome and is responsible for killing more than a million bats. This is especially concerning to farmers because healthy bat populations eat and control insects.

The increase in fungal illnesses is likely due to several factors. These include an increase in global temperatures, global travel, and global food production practices. Fungi love moist, warm conditions. Because they can form durable spores and reproduce asexually, fungi can also adapt quickly to changes on our warming, increasingly globalized planet. Warm-blooded animals like ourselves don't have these adaptive advantages.[1-3]

These characteristics also make fungi more deadly than bacteria when they get out-of-control. Robert Kremer, a USDA microbiologist who has become familiar with out-of-control fungi, has published research that sheds some light on my father's mycotoxicosis problems.

Kremer, who has compared glyphosate's effects on soil and ecosystem health, noticed that more fungi grow in areas sprayed with glyphosate-based herbicides. Specifically, he demonstrated that glyphosate affects the number of beneficial fungi and microorganisms growing around soy bean roots. He also noted that many plants treated with glyphosate mixtures have smaller root masses and fewer root nodules than other crops. Healthy roots and nodules are important because they help plants absorb nutrients and water.[4,5]

Kremer also noticed that crops only absorb a fraction of the herbicides sprayed on them. Glyphosate-based herbicides can move through the plant and out the roots where they exert additional antibiotic-herbicide properties. Here, as we mentioned in Diagnosis Two, glyphosate continues to tie up nutrients and cling to soil particles. According to Kremer, soil residues can accumulate to be several times the amounts found after application. These residues can later influence fungal growth in farm fields while reducing plants' abilities to retain nutrients and water.[4,5]

Glyphosate's influence on fusarium levels is significant in and of itself. In 2009, Kremer demonstrated repeatedly that as fusarium mold populations grow, other important soil organisms decline. In culture, mold was visibly present to a greater degree on plant roots treated with glyphosate formulations. While these effects were not always immediately apparent, Kremer reported that they usually appeared within a week.[6-8]

The link Kremer established between glyphosate and fungal diseases is not new. In fact, similar links were observed more than twenty years ago. Fusarium colonizations can cause several devastating diseases for American crop farmers including Sudden Death Syndrome in soybeans, Fusarium Head Blight in wheat, and Fusarium Wilt and Root Rot in cereals. Such diseases are becoming more severe in areas around the world where non-modified crop varieties are planted in fields previously used for GR (glyphosate resistant) crops. Past Canadian studies, for example, have shown that previous glyphosate applications contribute to high levels of fusarium head blight. Glyphosate applications somehow alter plant metabolism increasing a plant's susceptibility to mycotoxin induced diseases.[9]

Unfortunately, even though fusarium molds can colonize the root, the crown, and the kernal of a plant, most studies have been directed at the grain because the grain is what animals and humans eat. This could be problematic when farmers use the rest of the fusarium-contaminated plant for bedding.[9] My father suspects that glyphosate treated straw and corn stalks used for animal bedding could contribute to rising levels of mycotoxicosis because mycotoxin levels in many bedding sources exceed clinically significant levels for infertility and toxicity.

Curtailing mycotoxins in grains and animal bedding is necessary to prevent increased mycotoxin issues in our food supply. Unfortunately, the globalization of current cultivation practices makes doing so difficult. While technological improvements in mechanization and genetics have allowed farmers to plant soybeans and corn closer together, these same advances have compromised plant and animal health. The dense canopy of plants covering fields lends to greater moisture and heat retention at the soil level, creating perfect conditions for fungal growth, especially during wet years.

Thanks to these improved methods, farmers don't have to till their fields or rotate their crops as often, and fungus can build up in soils. The mycotoxins these molds produce gradually accumulate at the base of the human and animal food chain.

To address fungal problems, the pesticide industry began to market crop fungicides. This market has grown considerably in the last decade. Prior to 2002, fungicide use on corn was so uncommon that it wasn't even reported in the National Pesticide Use Database.[10] Today fungicides are used widely as growth enhancers on corn crops throughout corn producing regions. This happens despite the fact that the efficacy of such applications is still debated. Most fungi affect lower regions of corn, bean, and wheat crops: regions fungicides wouldn't reach anyway.

Of course, the risk that these fungicides will generate new fungicide resistant molds is high. These risks increase when fungicides are used to prevent and treat non-fungal diseases. Some bacterial diseases, like Goss' Wilt look like fungal infections, but are not. It would be very easy for any farmer to think that fungicide might be needed to reduce the prevalence of Goss' Wilt. Unfortunately, misplaced fungicide applications increase plant infections by reducing beneficial organisms in farm fields.[11]

Fortunately, farmers don't need to do anything new to confront growing fungal problems. In fact, some scientists are recommending that American farmers return to something old. According to these scientists, the best way to combat pathogenic fungi is to diversify crop systems and use better managed crop rotations.[3-8] Fungi need air to grow. No-till systems promoted by herbicide dependent, GM cropping systems only encourage fungal growth. The organic debris left to rot on the soil's surface is an ideal habitat for fungal colonies. These habitats extend for hundreds of acres and are seldom rotated when corn production is incentivized year after year. Farmers who used crop rotation and strip cropping years ago were already good at addressing fungal issues before companies convinced them they needed new fungicides.

We have effectively created a global food production system in which both pathogens and detrimental ideas can easily spread and multiply. Scientists with powerful reputations trick us into believing that if other nations don't accept modern technological advances, we will never be able to feed the hungry. They often do this while implying that those opposed to GM crops and their chemical counterparts don't fully appreciate the struggles endured by the world's poorest inhabitants. Unfortunately, these advances don't often reach the poorest inhabitants in ways that are truly helpful or sustainable.

In "Feedstuffs: The Weekly Newspaper for Agribusiness," Dr. Richard Raymond routinely expresses such "viewpoints" in his column. In one editorial entitled, "How Far Back to Ag Basics Do We Go," Raymond chastises ignorant Americans who don't appreciate what ag advances have done for modern society. He suggests that "if we go back to the "good old days," prior to such advances, "when all cattle were grass fed, the U.S. would need an increase in pastureland equal to the size of Montana and New York." We would also have to rely on more people to grow our food, and be willing to pay as much as we did in the 1930s when Americans spent 25% of their disposable income on food compared to the 5-10% we spend today.[12]

I would have to argue that such changes, albeit in modified amounts, wouldn't be such a bad thing. By returning 20% of the land currently used for monocultures back into rotational grazing and using another 10% to establish diverse vegetable and fruit farms, we could create healthier, more diverse ecosystems. What's more? We would still have plenty of corn and bean territory left for grain production.

Of course, we would need to depend less on corn biofuels to make such transitions possible. But other scientists have already pointed out that more alternative biofuel technologies, including the use of some combination of perennial biofuels, electric car models, and wind and solar power, could better serve our goal to maximize domestic transportation potential anyway.

Raymound's second point, that more people would be needed to work diversified farms, also seems far from problematic. At present, many Americans can't afford healthy food products because they don't have decent jobs. If we create meaningful jobs and a larger supply of healthy fruits and vegetables simultaneously, we can address two national problems at once.

Unfortunately, consumers will need to reduce average meat consumption to make such transitions possible. Not only is meat undervalued, Americans eat way too much of it. Today we live in a country where we can eat five burgers for five dollars all in one sitting even though we have fewer natural resources per person than we did many years ago. These prices are neither healthy for the consumer eating the burgers, nor the animals being raised so quickly and carelessly to produce them.

Livestock farmers might be alarmed to hear me suggest such a thing. How can an animal ag proponent discourage consumption of animal products? Here I need to clarify: while I *am* suggesting that Americans eat less meat per person, I *am not* suggesting that farmers make fewer dollars per product sold. If we hope to retain the greatest number of animal farms overall, we will need to add value and care to each farm animal.

Doing so will be difficult but worthwhile. To create demand for more value-added animal products, American producers need to feel confident that they can stay competitive with foreign food importers.

By limiting animal imports, we can enhance global food security and the regionalization of food ecosystems. Prices should better reflect the raw inputs and hard work required to raise healthy animals for consumption.

If food systems were better regionalized and those who could afford it spent 5% more on their food budgets each year, I probably wouldn't have as many students eating zero to three servings of green, leafy vegetables each week. Today many of the same students that eat very few fresh vegetables also have expensive cell phones, monthly manicures, and perfect adolescent complexions. It seems our policies have been teaching the middle class to put greater value on non-essential items and less value on essential nutritional products for some time.

If we had more vegetable and fruit farms, we might also have more healthy, fresh food items to donate to area food banks to help feed our nation's poor. Every citizen who helps support higher quality domestic food products is also helping to raise our mainstream expectations. Hopefully, over time these expectations will translate into lower fresh food costs for everyone. Today the opposite is often true: raw, whole foods often cost more than heavily processed foodstuffs.

Still we continue to incentivize cheap meat production and complain when we can't afford to consume 8-12 oz of meat per day, even though this isn't recommended. If organic vegetable costs were leveled out and new labeling laws were put in place, we should, as a nation be able to encourage a market that has more affordable greens to offset increased costs associated with value-added meat and dairy products.

Industrial ag proponents often completely ignore discussion of more ecologically friendly ag systems that could be useful in meeting these goals. These technologies have developed alongside gene transfer technologies and confinement operations. I would like to think that most Americans consider more intensely diversified ag systems *advancements* as well!

Nonetheless, these are not the advancements being encouraged around the world. During an era in which colonization has been frowned upon, we have been using one type of ag advancement to effectively replace skills and trades associated with diverse, regionalized systems elsewhere. A drive for profit, minus any thought as to how agbiotech practices affect local economies, is feeding the true root of several modern ag ailments, including increasing incidences of mycotoxicosis. In the following paragraphs, I explore how the spread of pathogenic ideas can affect several countries' diets at once.

One recent project promising to confront "famine and hunger in an effort to break the vicious cycle of poverty" fits this description rather nicely. According to the project's founder, Bruce Rastetter, AgriSol Energy's project would "help the people of Tanzania create a better life for themselves and their communities."[13] Since the project began, concerned citizens, politicians, and reporters have argued that AgriSol's project would actually do just the opposite.

How does an American agribusiness reduce hunger and poverty abroad? It decides to introduce modern ag technologies and developments to "global locations that have attractive natural resources." Its Tanzanian project will "transform food production in Tanzania from subsistence farming, with low productivity and under-utilized land resources, to commercial farming that can help foster mature domestic markets and value-added businesses." It will do so in a way that creates *sustainable* change for the region. According to AgriSol's website, project participants collaborated with academic institutions and the Tanzanian government. Their impressive goal was to create a *private-public-academic partnership.*[14]

While this all sounds wonderfully collaborative and beneficial, there are several reasons concerned citizens doubted the project's overall effectiveness. Shortly after AgriSol's investment partner announced its decision to spend $350 million on foreign farmland, AgriSol announced its plans for 800,000 acres of African land.[15,16] This land would be the site of new commercial livestock and grain operations. AgriSol would essentially help Tanzanians adopt GM crop and chemical practices inviting foreign agribusinesses into new markets.[15] Unfortunately, a government and people who were largely unprepared to regulate, let alone benefit from such systems, approved the project—a project that promised to benefit sponsoring multinational agribusinesses just as much as anyone else.[13-15]

While the project did involve some Tanzanians, reporters and organizations pointed out that AgriSol used manipulative rhetoric to reap hundreds of millions of dollars while offering to pay Tanzanians less than a dollar per acre for their land! In the process, approximately 160,000 African refugees would be displaced![15]

Several groups have called AgriSol's philanthropic investment an "African land grab" in disguise. They've pointed out that the company's pursuits could draw government resources away from area residents and create environmental damage associated with modern ag practices. Several Iowans, Tanzanians, and non-profits actually banded together to raise awareness about the project, accusing Rastetter and his partners of *shameful ethics violations*.[15,17]

In response to such criticisms, Rastetter wrote an op-ed piece to the Des Moines Register and posted it on AgriSol's website. In his letter, Rastetter assured Iowans that AgriSol "is deeply committed to the local communities" and "would never help or advocate for the removal of any person from land for one of [its] projects."[13] According to Ratstetter, the government and people of Tanzania welcomed the project.

Interestingly, an earlier televised report by Dan Rather revealed that very few local Tanzanians knew about the project. This included interviewed village officials. Later, AgriSol claimed that the government was already planning to resettle area refugees living in *uninhabited* sites. The Oakland Institute, on the other hand, found that evacuation of refugee populations was required for the project and that AgriSol communicated this clearly to governing authorities.[15] Provided that thousands of refuges lived in these sites, it's difficult to imagine how AgriSol's project could employ the people leaving this area or the individuals removed from ag in general.

Like leaders of other philanthropic "feed-the-world" projects, Rastetter has powerful connections. In addition to heading Summit Farms, the Iowa farm operation that oversees AgriSol, Rastetter served as CEO of Pharos Ag, a partner in the Rural American Fund (a Chicago based equity fund investing in agribusiness), former CEO of Heartland Pork Enterprises (a vertically integrated swine operation that declared bankruptcy), and former CEO of Hawkeye Energy Holdings (a major ethanol producer that also declared bankruptcy).[18]

These pursuits made Rastetter a very wealthy man. As a result, he's been able to contribute large sums of money to political campaigns and Iowa State University, a land grant institution specializing in ag research.[18,19] He used some of this money to endow the "Bruce Rastetter Chair of Agricultural Entrepreneurship" at Iowa State. Following this gift, Iowa State University advised Rastetter's AgriSol project. The college offered to help with local outreach, research, and outgrower programs in much the same way ag extension offices currently assist Iowan farmers. Readers need to be reminded here that most of these services are offered with an agribusiness bent.[18-20]

To distance itself from controversy, Iowa State University later announced that it would stop advising AgriSol due to misunderstandings about its involvement. Such misunderstandings prompted Iowa State students, residents, and legislators to gather for a forum entitled "After AgriSol: Defining a University's Ethics and Interests in a Corporate World." Forum members discussed land grant university roles as they pertain to public-private-academic partnerships. Those gathering seemed particularly concerned that their land-grant college was wiling to help develop commercial farms on refugee occupied lands.[18-20]

AgriSol has since postponed its plans to develop refugee occupied lands and claims that it is only moving forward on projects in uninhabited areas. Still, many wonder about its involvement there, especially in light of an increasing amount of research that shows that more diverse, alternative production systems are better options for such regions.

If Africans own the land, rather than outside investors, the money will be retained in their communities. If international conglomerates own the land, a portion of the profit will leave with them. To create lasting food security, one must consider models that work well within Tanzanian systems, rather than buying up land and teaching a nation how to feed its people the American way.

Agriculture is place-specific and needs to be matched to local resources. Projects such as these introduce foreign expectations, energy needs, and genes into ecological and political systems that are not anticipating these things. These systems then need additional support to deal with regulatory and environmental challenges as they become apparent.

Such exploits aren't all that different from what the colonialists did when they moved in on the Americas. Native Americans weren't ready to handle our guns, germs and steel and we literally wiped their cultures out—a massive genocide that effectively eradicated more than 90% of the Native American population.

Today we could do this with mycotoxins and not even know it. How will our genetically modified seeds and chemicals interact in foreign ecosystems? What field trials, if any, are done before investors purchase lands and market such ideas?

I hope that Tanzanians are better prepared for our gene guns and plant diseases than we were—that they can foresee which native plants will be affected by cross-pollination and which insects will thrive on patented commercial crops. I wish that I could come up with a better way to translate America's two-headed language before foreigners adopt more monoculture, row cropping practices. Meanwhile, external interest groups take advantage of land that isn't really theirs to take advantage of in the first place.

Some wonderful reporting by ag columnist Alan Guebert highlights Rastetter's controversial connections with powerful figures who have all benefited from global philanthropic ag investments. Guebert refers to this concentration of powers as the "Good Old Farm boy Network." He goes on to explain how such networks push their agendas on foreign parties.[20]

Unfortunately, developing nations have already learned too many anticultural lessons from *Good Old Farm Boys* in the past. International trade and economic agreements have often pushed these nations into difficult decisions to accept genetically modified foods. According to Patrick Mulvaney, Senior Policy Advisor to Practical Action in Kenya,

> Those with power, particularly the United States, have used hunger as justification for trade supremacy and the promotion of proprietary genetically-modified (GM) crops owned by northern multinational corporations much to the delight of pro-GM advocates. Countries and their peoples who legitimately resist the consumption of GM grain and seed are under intense diplomatic threat of denial of food aid in times of crisis. United Nations agencies and US private voluntary organizations are complicit in this process...[21]

Mulvaney goes on to explore how countries like Sudan, Angola and Zambia tried to stand up to GM food aide. Residents of Angola, for example, supported GM food aide bans on the basis that it would reduce their genetic resources. The resources they hoped to protect included approximately 800 different types of corn.[21] When corn introduced as food-aide replaces or cross-pollinates these varieties, these species, like many American corn varieties, are lost to regional farmers.

Unfortunately, obtaining non-GM feed sources isn't getting easier for anyone. Agribusiness publications report increasing numbers of international acquisitions and mergers that threaten to squeeze out local ecotypes. Dupont, for example, recently announced its intent to acquire a majority share of the South African seed company Pannar Seed Ltd. According to a *Feedstuffs* article describing this acquisition, "Africa represents a significant opportunity for maize production, with 80 million acres available for cultivation."[22] This land, in the eyes of foreign investors, is simply up for grabs!

As a condition to receive approval, agribusinesses are establishing research centers to bring advancements to Africa. But as African countries approve patented seed stocks, will they maintain their rights to grow their own seeds? First world, American farmers, familiar with American corporate language aren't able to do as much.

According to a 2010 publication by the Center for Strategic and International Studies (CSIS), several needs should be met before foreign countries adopt GM crops. Nations hoping to retain control of their food supplies should have an open and informed political debate, a holistic approach to boosting ag productivity, regional policies to address commercial ag systems, and a public campaign to address human safety, environmental contamination, and ethical considerations.[23,24]Indeed, they are entitled to many of the things we never gave ourselves!

Developing countries also need to keep a close eye on trade relations, being particularly vigilant to avoid "dumped" seed commodities. A "dumped" product is any good that is sold for less than fair market value. When ag commodities like corn and beans are dumped into foreign markets, these artificially cheap goods can drive traditional farmers out of business.

Researchers at the Woodrow Wilson Center have already analyzed how free trade relations, which sometimes encourage dumping, have played out between the United States and Mexico. They came to the conclusion that Mexican corn farmers suffered numerous losses during the nine-year period studied. They reported that:

> U.S. exports increased 413%, arrived at prices 19% below production costs, and real producer prices in Mexico declined 66%. We estimated losses to Mexican corn farmers of $6.6 billion over the nine-year period, over $700 million per year. These losses amount to $99/hectare per year, a crushing blow to struggling smallholders.[25]

The authors of this report went on to say that dumped corn products led to "crushing losses for Mexico. An estimated 2.3 million people have left agriculture in a country desperate for livelihoods….And food dependency has risen dramatically, which cost Mexico dearly when commodity prices spiked in 2006-8."[25]

United States farmers have simultaneously relied on subsidies and insurances at home, while frowning on such practices elsewhere. Our politics have undermined agricultural free trade. Other nations are unable to compete with our crop prices unless they accept our messed up systems themselves. In this context, *free* trade systems that were originally designed to reduce price distortions and promote economic justice are doing just the opposite. These free trade agreements have done little to recognize the place-based nature of agriculture around the world. Just as one person's educational needs and goals may differ from another's, agriculture is incredibly specific.

As a teacher, one of the most difficult aspects of my job is that I am supposed to regularly address each student's Individualized Education Plan (IEP) or 504 plan. These plans are legally binding documents outlining lesson modifications that a student needs in order to receive a *fair* public education according to the law. Students might have one of these plans to address struggles related to a form of autism, a serious case of ADHD, a learning disability, or a behavioral disorder.

In an ideal world, I could address every recommendation outlined in this kind of plan. Recommendations include modifications to student work such as highlighting key concepts, enlarging worksheet print, providing extra visuals, and modifying test length. The hope is that by making these changes, each student with an IEP can be successful in the classroom. This is not very difficult when 1/10th of the students in a classroom has a special set of modifications. However, it starts to feel impossible when several students with numerous modification requirements, are mixed into rooms with students performing at or above grade level.

Middle school students are attentive to IEP's. If they see that Johnny gets an extra study-guide for a test or a word bank because he has a reading disability, they often want to know why they aren't getting one as well. *It's unfair! Why don't I get one?* Teachers, in general, try to be discreet about student modifications to protect both the child and ourselves. We also frequently remind students that treating people fairly doesn't always mean treating them equally, even though we would otherwise encourage equal treatment.

A student's IEP or 504 plan recognizes that there is a larger system—a system of free, fair public education for all, but that for some disadvantaged students, the system will never seem fair or free. It's a valuable lesson that we need to carry over into our discussion about ag trade policies. In our efforts to rapidly globalize we have forgotten that we all bring different resources and needs to the table. Any single, globalized regulatory system, encouraging the patented sale of pesticide resistant, genetically modified crops, doesn't fit all.

In theory international free trade, where participating governments do nothing to manipulate import and export prices, works. This system is "free" of tariffs and subsidies, as well as subsidized crop production. Prices emerge from supply and demand and are largely determined by a nation's resources as opposed to government interventions. By conservative estimates, a free trade system that is *truly* free should increase national incomes by billions of dollars.

In the real world, prices are distorted for a variety of reasons and free trade doesn't live up to these ideals. Since 2001, the World Trade Organization has hosted a series of negotiations called Doha Rounds to improve global free trade. Participating countries have discussed a wide range of issues, but seem particularly stuck when it comes to agriculture. Disagreements between developed and developing nations continue to impede resolutions. As a result, no one seems very sure as to what will become of future Doha conversations.

Agriculture is the most controversial subject at Doha because many people depend on agriculture for their livelihoods. In some third world countries, more than 70% of a nation's inhabitants depend on ag for their main sources of income. Bad decisions stand to jeopardize many futures.

In 2007, a trade compromise at the Doha rounds looked pretty promising. The U.S. announced that it would reduce its farm subsidies if Brazil and India dropped other objections in the round. The round collapsed, however, when India and China maintained that U.S. subsidy concessions were not enough. In addition to these problems, there were debates between India and the United States over something called the special safeguard mechanism (SSM), a measure designed to protect poor farmers by allowing developing nations to impose tariffs on important ag goods. The idea was that these tariffs would protect local farmers from import surges and price collapses.

United States representatives argued that the thresholds for implementing differential treatment were too low. Others argued that wealthier nations were sacrificing overall goals to satisfy national interests. To put this in perspective, the amount of U.S. subsidies alone, at that time, was larger than several developing countries' entire budgets! While it seemed to take a lot for the U.S. to arrive at a 3.2 billion dollar concession, the United States still walked away looking like a spoiled child.

The same companies sponsoring "feed the world" partnerships pressure politicians to distort trade practices in a way that robs third-world individuals of their abilities to produce and obtain nourishing food. And despite all of our twenty-first century, well-educated awareness, the world's freer markets aren't getting any freer. The truth, in this case, is that free markets minus regional plans will never succeed. For these reasons, I have come to oppose agricultural free trade.

While true free trade might result in huge net gains overall, the wealth that is currently accumulating from these systems mostly serves those who are already well established. [26-31]Agriculture is special and should be treated as such. It serves one of humankind's most basic needs. If our true intentions are to help developing nations, we need to encourage some protectionist policies to ensure traditional jobs and businesses remain viable. More protectionist policies, like taxing imports on foreign dairy and corn products, would help discourage trade-distorting practices like dumping, while improving food security for various countries around the globe.

When philanthropists gift poor countries genetically engineered foods and seed technologies, they are also gifting these countries new genes and chemicals in need of regulation. This seems such a shame because we, ourselves, have trouble assessing and regulating these seeds and chemicals. Such actions might be excusable if there weren't other options. But think tanks have already proposed numerous alternatives that rely less on commercial agriculture.

One alternative system, called "Save and Grow," actually has the support of the U.N.'s Food and Agriculture Organization (FAO). This system would feed the world by *intensifying sustainable, regional agriculture systems on a global scale*.[32] The director general of the U.N., Jose Graziano da Silva, believes that hunger reduction and the environment are *irrevocably connected* and that development cannot be sustainable when "nearly one out of every seven men, women and children are left behind, victims of undernourishment."[32]

Proposed systems like Save and Grow could help all global citizens conserve precious energy and water resources needed to grow healthy foods. Gene transfer events designed to rely on pesticide-heavy, monoculture systems are not necessary to achieve these ends.

While the FAO does not openly condemn GM technologies, it does discourage heavy reliance on herbicide tolerant crops. The authors of the Save and Grow plan also explain how Integrated Pest Management (IPM) strategies, which incorporate more ecological pest control systems, can reduce current pesticide use. The plan acknowledges that over-use of pesticides

> contributes to a vicious cycle of resistance in pests, which leads to further investment in pesticide development but little change in crop losses to pests, which are estimated today at 30 to 40 percent, similar to those of 50 years ago. As a result, induced pest outbreaks, caused by inappropriate pesticide use, have increased.
>
> Excessive use of pesticide also exposes farmers to serious health risks and has negative consequences for the environment, and sometimes for crop yields. Often less than one percent of pesticides applied actually reach a target pest organism; the rest contaminates the air, soil and water. [33]

All citizens could benefit from better understandings of items incorporated in the Save and Grow plan—including understandings of how organisms like fungi, bacteria, and other supposed "pests" benefit agriculture in the long-term.

Unfortunately, most children aren't growing up with these understandings. Instead, American children watch companies copy and paste entire ag models into foreign lands without officially "colonizing" these lands first. Children hoping to explore the African savannah may soon find themselves driving through thousands of acres of GM corn and soybeans just to arrive at another American fast food joint. Here they can enjoy African corn-fed burgers. Perhaps some of the technological improvements supported by *private-public-academic partnerships*, will also help develop "improved transportation networks" through these savannahs so GM grains can "more efficiently feed the world." These interstates will completely disrupt remaining big game ecosystems as well as the cultural and economic values these resources once held. Perhaps each of these continents can then fund research to address new pandemic mycotoxin issues that used to be limited to certain regions?

We need to jump off this highway to Ubiquity and get back on the path to Somewhere real quick. Dr. Seuss, who is perhaps my favorite doctor, didn't marvel at all of the world's sameness, he marveled at all of the "places you'll go!"

While increases in travel and temperature encourage new and unusual fungal infections, the international exchange of food products increases the likelihood that a food safety epidemic will occur. A pathogen that is successful in one system will have equal success in carbon copies of this system elsewhere!

Global epidemiology concerns create regulatory nightmares that end up costing the world even more unnecessary dollars. Some of the food safety questions we currently debate include: How should internationally traded meat and vegetables be labeled? Should countries have to report every chemical and antibiotic used? How can we best trace contaminated meat and vegetable products when they end up in a country that doesn't regulate them? Furthermore, when an epidemic does occur, who should address the problem: the importer or the exporter? These questions are difficult to answer because we didn't establish accountability in the first place.

Because we have already been so "successful" at thrusting our policies and practices on others, we may have to work backward a bit to undo some of the damage that has already been done. Specifically, we will need to make it more difficult for commercial companies to assume global monopolies by better monitoring their educational campaigns and international seed patents. There is no reason a profitable venture like Dupont or Monsanto needs to own hundreds of thousands of acres and seed varieties in third world nations in addition to the seeds and acres they already control.

Groups who truly care about global food security and sustainability will focus on these goals rather than continuing to export models and rhetoric that discourage biodiversity. My father actually listens to clients vent about trade issues all of the time. These clients rightfully wonder how our ability to "freely" import beef or soy from Brazil will ever benefit farmers in the United States. As more foreign feed products are allowed into our markets, farmers at home will be pressed to cut costs, inflicting even more stress on animals and systems that are already taxed. These trade practices ultimately affect the health of crops and animals raised for human consumption

My father, like many farmers, is learning these lessons the hard way. Treating mycotoxin issues is very complicated and can take months to resolve, even after the contaminated feed sources have been identified. Fungal colonies can be long-lived and resistant to treatment, in which case, an entirely new population of livestock animals may be necessary before true change is observed.

DIAGNOSIS 9: MULBERRY HEART SYNDROME

In addition to being one of the world's most intelligent mammals, pigs are, by nature, very clean, social creatures. They can respond to their name in a matter of weeks and pick up tricks more quickly than most dogs. They can also use their grunts to communicate and, when allowed, form relationships with other pigs and people. My father is not the only farm kid who has fond memories of his childhood pal: a pet pig. We like to teasingly remind him that this is why he is so much better at communicating with animals than people!

For years, scientists have known that pigs can recognize nutrient deficiencies in their diets. When options are provided, pigs can select protein mixtures most appropriate for their age and nutritional needs. A wild pig is an omnivore and enjoys eating nuts, roots, fruits, mushrooms, bugs, rabbits, and, occasionally, dead animals. They use their snouts to root around in the soil obtaining some nutrients directly from the ground. Post-weaning, industrial farm pigs, on the other hand, are normally provided a single grain and protein based diet for each of their growing and finishing phases. This diet is supplemented with nutrient inputs that may or may not be matched to its needs.

Tail-biting and aggressive behaviors have been linked to nutrient deficiencies. When pigs are salt, phosphorus, iron, copper or calcium deficient, exploratory chewing behaviors escalate. Pigs then begin tail biting and nipping, sometimes wounding other pigs. Nutrients in the blood, such as salt and protein, can increase a nutrient-deprived pig's appetite for more, especially when the pig already feels agitated in close quarters or undesirable conditions. Obviously, such behaviors can make managing pigs difficult while inducing damage to individuals in a confinement unit.

I imagine hog producers with such problems might feel somewhat like teachers attempting to manage high-density classrooms with five or more hyperactive students. Give me three, even four students with ADHD in a 30-student classroom and I feel fine. In fact, I thrive on a little enthusiasm. Entrust more children with such a diagnosis to my care and I begin to feel like I'm running a three-ringed circus.

To prevent such behaviors, many farmers supplement their grain-based feedstuffs with nutrient products. One commonly used pig supplement is mined phosphorus (dicalcium phosphate), a nutrient involved in DNA repair, neurotransmission, cell membrane function, and bone formation. Pigs, unlike ruminants, are largely unable to digest grain phosphorus without supplements. This is because they don't have the same beneficial enzymes needed to break a plant's phosphoric acids down into usable forms. In the last decade, farmers have replaced some of the mined phosphorus supplements with synthetic phytase, a man made enzyme that helps make plant phosphorus more available to pigs.

By replacing some of the mined phosphorus in pig feed with phytase supplements, farmers have been able to save significant amounts of money. Because phytase costs the producer 75 cents less per pig than traditional phosphorus supplements, savings really add up over the course of a pig's lifetime if you own lots of pigs. A producer operating a 200,000 pig unit can save roughly $150,000 annually by switching from mined phosphorus sources to phytase supplements.

Historically, supplementation of mined phosphorus inputs led to phosphorus buildup in pig manure. This was problematic even before farmers used phytase supplements because phosphorus can run off fields into neighboring streams and groundwater causing environmental problems.[1] Excess phosphorus levels are one source of nutrient pollution damaging our streams, rivers, and gulfs.

Today the phytase engineered to reduce phosphorus inputs can actually exacerbate phosphorus pollution. Before phytase was made a more stable compound a few years ago, the phytate-phosphorus complex excreted in hog manure stayed in the soil unless heavy rains pulled it away with soil loss. Now, the manufactured phytase is so stable excreted phytase remains active in the manure pit and soil even after the manure is spread as fertilizer.

Once it has reached the soil, manufactured phytase continues to make soil phosphorus more water-soluble increasing the likelihood that phosphorus can leach away (like nitrogen) in rainy, wet conditions. Soil phosphorus tests have dropped to 5 ppm/acre in some fields heavily fertilized with swine manure. This indicates that the phosphorus in hog manure is running off farm fields in greater quantities than we've seen in the past.[2]

Algae and cyanobacteria thrive in high phosphorus conditions and consume a disproportionate amount of available oxygen in our nation's water bodies. These algal blooms deprive fish populations of oxygen. When fish start to die, fishermen lose their sources of income.

High nutrient levels have significantly reduced aquatic biodiversity in areas like the Gulf of Mexico's Dead Zone. Since its discovery approximately four decades ago, the Dead Zone has steadily increased in size. The excess nitrogen and phosphorus running down the Mississippi has killed fish and shrimp and caused previously employed workers to emigrate elsewhere. Today, experts estimate that the Dead Zone is approximately 6,700 square miles, covering an area larger than the state of Connecticut. This zone costs regional fisheries around $2.8 billion per year.[2] As fish move further out to sea, fishing becomes even more expensive and difficult.

Certainly, other urban and rural processes contribute to ecological dead zones. Human sewage, lawn-care products, industrial byproducts, and row-crop fertilizers are equally damaging, if not more so, than animal feedstuffs. Nonetheless, reducing excess inputs in all sectors is necessary to manage damaged areas. Unfortunately, in the world of animal supplements, farmer and consumer savings often come with overlooked consumer and environmental costs.

Rather than reevaluating phosphorus sources or rethinking the way we recycle pig manure in site-specific contexts, some scientists hope to reduce pigs' environmental impacts by introducing the *enviropig,* a pig engineered to produce the enzyme phytase.[4] The idea is being marketed as a need that, quite frankly, has largely been created by modern convention. This need will likely come with additional, difficult to assess costs.

Phosphorus is only one of several nutrient supplements my father is currently concerned about. Unfortunately there is a common understanding that synthetic supplements are equivalent to their naturally growing counterparts. Not only is this not necessarily the case, synthetic nutrient supplements can cause unnecessary damage in human and animal systems. In this chapter, I give you a glimpse of how complex, multi-supplement interactions affect animal health.

My father has learned to appreciate the complexities of nutrient/tissue interactions while working with vitamin E deficient pigs and cattle. Vitamin E is a powerful fat-soluble antioxidant that scavenges for free radicals that can damage various tissues. Humans take vitamin E supplements as part of age-defying skin regimens and antioxidant products designed to destroy cancer-causing agents in the body. Vitamin E also helps body cells regulate gene expression. When animals don't have enough vitamin E, they have a harder time making amino acids, vitamin C, and colostrum, the extra healthy milk produced immediately following birth. In fact, vitamin E is necessary for optimal circulatory, nervous, muscular, reproductive, and immune system activities.

Given suitable conditions, pigs can obtain enough natural vitamin E from various feed-stuffs including grains, green plants, and vegetable oils. However, in many current industrial farm diets, vitamin E is not present in high enough concentrations to counter additional stress placed on confined animals. While most farmers use vitamin E supplements, vitamin E deficiencies continue to be problematic because they aren't always bioavailable even if the E is there chemically.

Several feed practices complicate vitamin E absorption in an animal's body. One major problem on industrial farms is that some of the poly-unsaturated fats in modern feed byproducts and mixtures actually destroy the vitamin E before it is absorbed.

To understand exactly what's going on, you need to know a bit about vitamin E chemistry. There are several forms of vitamin E molecules used by the body. These forms are broken down into two subgroups called tocopherols and tocotrienols. Tocopherols and tocotrienols have slightly different molecular structures accounting for their different functions.[5] To make matters more complicated, each type appears in four three-dimensional arrangements, typically indicated with a prefix.

Historically, most scientists thought that d-alpha tocopherol, which can be found in foods like greens, sunflower oil, and safflower oil, was the most biologically active form of vitamin E. Confinement pigs and humans fed a diet high in corn and soybean oils consume much more y-tocopherol than d-alpha-tocopherol. Not only is this form less active, it is often destroyed during feed processing and storage. For this reason, many farmers now supplement their livestock herds with d-alpha-tocopherol.

In the last decade, scientists have learned that tocotrienols, which are less widely available, may be much better at combating oxidative stress associated with cancer, high cholesterol, and cardiovascular disease.[6-12]

Tocotrienols can be found in natural sources like nuts, barley, and wheat germ.

Misunderstandings regarding the ways tocopherols and tocotrienols interact has led to a misuse of synthetic vitamin E supplements. Synthetic vitamin E is different from most naturally occurring vitamin E sources because it contains both naturally occurring d-alpha tocopherol and its three-dimensional mirror image. The mirror image is identical in structure with the exception that three parts of the molecule bend in a different direction in three-dimensional space. The shape of these synthetic molecules influences the way the body recognizes and absorbs the purchased supplement. For this reason, all natural d-alpha-tocopherol extracted from corn oil, is twice as active in pigs as synthetic versions containing its inactive mirror image. Unfortunately, synthetic versions of pure "d" tocopherols and tocotrienols have not been commonly available.

The synthetic supplements farmers have been using also contain varying ratios of tocopherols to tocotrienols. Negligible amounts of tocotrienol may be problematic because a variety of studies have shown that synthetic tocopherol can actually interfere with tocotrienol's benefits.[6-12] In these instances, low doses of alpha tocopherol have been found to be beneficial, while high doses appear to increase cholesterol levels reducing tocotrienol's ability to fight cancer-causing cells. Supplements with high tocotrienol to tocopherol ratios may be much better for humans and animals than those with low ratios. Unfortunately, because the opposite is usually true (tocopherol levels are typically higher), many scientists have questioned the effectiveness of widely marketed supplements. In fact, scientists have reported that over supplementation with vitamin E, at levels similar to those found in widely marketed human and anmal supplements, may increase the risks of several ailments including osteoporosis.[13-19]

My father is not a synthetic vitamin expert. However, during his thirty-nine years of experience working with cattle and hogs, he has observed that nothing beats natural vitamin E from sun-cured hay that is not too over-cooked or rained on. The best haylage and corn silage in the world often has no natural vitamin E left in it. The fermentation process that helps preserve these feed products destroys them. While some forms of vitamin E supplementation do reduce herd problems, he is not convinced that vitamin E supplements are working as they should. My father has actually seen an increase in vitamin E deficiency issues during the last two decades.

The quality of supplements used makes a big difference on the farm. In the last several decades, the firms handling trace mineral and vitamin products have consolidated internationally. Today, most animal trace mineral and vitamin supplements that come under different brand names originate in the same areas of China and Southeast Asia. They are then sold and distributed to numerous feed companies throughout the United States where they are mixed with other feed products. Due to a lack of quality control and various shipping, handling, and storage issues, the end-quality of these products can be very hit or miss.

Vitamin E deficiency has been especially problematic in modern confinement units. When pigs are vitamin E deficient they can contract mulberry heart disease. One day a pig looks relatively fine. The next day it up and dies of a heart attack.

When a veterinarian posts a pig with mulberry heart to determine the cause of death, the heart looks like a mulberry because it has been over-worked and damaged. Fluid has accumulated around the heart, lungs, and abdomen, and muscular areas appear pale because they haven't gotten enough oxygen.

Pigs with mulberry heart disease ultimately die of heart attacks but there are several underlying stressors that lead to heart failure. While this would not be unusual if the pig had lived a long life, many of the pigs dying of mulberry heart disease in modern confinement facilities are only weeks old. Imagine a baby having a heart attack at just a few weeks old! To make matters worse, my father reports that mulberry heart disease used to be limited to rapidly growing piglets shortly after they were weaned off their mothers. Today, older fat hogs are dying of this strange Vitamin E disorder as well.

Why this is happening at any frequency should be a concern. When sows farrowing piglets are vitamin E deficient, the piglets born to these sows die suddenly after they're given routine iron injections. Their mothers' low vitamin E levels have already affected their abilities to absorb iron. This is because vitamin E is necessary to help piglets develop the appropriate number of heme molecules. A piglet needs heme molecules in its blood to bind free iron.[20-24] In fact, sometimes my father finds that the dead piglets' carcasses are literally the color of iron, an earthy brownish red. The injected iron has no place to go!

In a healthy pig, this iron would be used to make hemoglobin. Hemoglobin is necessary because it binds to oxygen and carbon dioxide and delivers these life-sustaining molecules to and from body tissues. Without oxygen your cells can't breathe and carry out metabolic processes even when your lungs inhale and exhale air.

When an animal, including a human, is low in iron content, we say he or she is anemic. Baby pig anemia has always been an issue in confinement litters because nursing pigs don't have access to the iron in the soil as they would if they were out on pasture. Even in the absence of a vitamin E deficiency, confinement piglets need 200 mg of iron in the first three days of life to get a healthy start. They use this iron to make hemoglobin. In fact, if you don't give confined pigs iron injections, they will be smaller and more prone to disease (like enteritis and wasting) for the remainder of their lives.[20] When a pig is vitamin E deficient, these iron injections are useless because the lack of vitamin E during development already interfered with the injected iron's docking system.

Because numerous nutrient and mineral interactions affect vitamin E and iron absorption, treating mulberry heart disease is rather complicated. Unfortunately, there are several other nutrients and minerals we have to pay attention to as well. Selenium, for example, is needed in minute quantities to ensure a pig doesn't develop mulberry heart. Like vitamin E, selenium is needed for proper cellular function and helps anti-oxidize harmful materials. In fact, when an animal doesn't eat enough plant matter containing selenium salts, it needs even more vitamin E to protect its cell membranes, DNA, and active enzyme sites.

On the flip side of the coin, when pigs ingest too much selenium, they can actually die of mulberry heart disease because the extra selenium acts as another heavy metal to oxidize vitamin E, destroying vitamin E's ability to convert harmful molecules into less damaging particles. If a piglet's mother is vitamin E deficient and her piglets are anemic, selenium in high amounts just creates one more unnecessary demand on the piglet's developing systems.

Making sure pigs are receiving appropriate amounts of iron, selenium, and Vitamin E is essential to prevent conditions like mulberry heart disease. When my father posts a piglet that has died of mulberry heart, he tries to determine which of the three is the biggest contributing factor. This way he can better advise his clients on how to change their animals' diets.

During the seventies and eighties, selenium was often the most problematic. Scientists learned that some areas of the United States have more selenium deficient soils than others and encouraged feed processors to add selenium to pre-mixed feeds. In 1974, the FDA allowed hog farmers to add selenium to animal feed mixtures at .1 ppm but over the next few years they raised this level to .3 ppm. As purchased feed mixtures became more common, feed supply companies started to adjust the levels of selenium in pig feeds before it was delivered to farms.[24]

Soils are selenium deficient in much of Northeastern Iowa. When the government first approved selenium use in animal feeds, many of my father's clients were already adding it to feed mixtures themselves. As a result, some animals received twice the amount of selenium needed when feed companies began adding selenium to pre-mixtures.

As I've already mentioned, selenium levels that are too high can be just as problematic as selenium levels that are too low. In fact, animals can die from selenium poisoning. It took my father years to convince some clients to stop adding extra selenium to their feed products because it wasn't always clear when and where it was being added. Ironically animals were suffering from selenium overdose in an area that is naturally very selenium deficient.

When the government took steps to reduce the amount of acid rain generated by coal power plants and diesel engines in the early eighties my father received another lesson about selenium, iron, and vitamin E interactions. Up until this point the sulfur dioxides emitted from diesel engines and smokestacks acted as a source of free sulfur for Northeastern Iowa's sulfur deficient soils. While acid rain can certainly cause many problems, the sulfur from these rains actually provided a benefit to animals eating low-sulfur forages in my father's practice.

When the sulfur from these rains became incorporated in area hay, the animals eating these sulfur-rich grasses had more sulfur available to make an important amino acid. Sulfur, as it turns out, is necessary to make the amino acid methionine. Methionine, in turn, is important for healthy milk production.

Animals without enough sulfur can develop very bad cases of milk fever, a postpartum condition in which cows becomes weak, lose their appetites, and, if left untreated, die of heart failure. Simultaneous changes in sulfur and selenium levels exacerbated these symptoms in several herds eating locally grown hay. As it turns out, sulfur and selenium bind to molecules in similar ways because these elements have similar structures. When selenium levels are too high, selenium will take the place of some sulfur atoms in sulfur containing compounds like methionine. Where ever methionine goes selenium follows. This is a problem for a cow or developing calf if the cow is already sensitive to elevated selenium levels in her diet.

So about a decade ago, when my father was simultaneously addressing several milk fever and selenium toxicity nightmares, he had to piece together a very complex nutrient interaction problem to solve herd health problems.

According to published research, selenium levels need to be 5-10 x higher than normal to cause poisoning. He could hardly blame the number of selenium toxicity issues he was seeing on feed overdoses alone. Animals out west eating twice the amount of selenium (on selenium rich soils) weren't suffering from milk fevers and selenium toxicities to the extent he was seeing in his practice. The selenium levels were also being impacted by the amount of sulfur in the environment.

In attempts to remedy his milk fever issues my father added magnesium sulfate (Epsom salts) and calcium sulfate to pregnant cows' diets. This did reduce the milk fevers but also caused healthy cows to develop diarrhea. It actually wasn't until farmers put calcium sulfate (gypsum) fertilizers back on the area's sulfur-deprived fields that he saw a true change in animal health. Sulfur levels eventually increased in the hay improving overall hay quality. Farmers got better yields and the hay contained more methionine, biotin, cysteine and thyamine needed for healthy protein production. The animals, in turn, could digest the sulfur in this hay without suffering the indigestive effects that arose when my father attempted to treat them with magnesium and calcium sulfate.

This was an important lesson for my father. He learned that is always better to have the nutrients and minerals incorporated in the feed crop than just added on top. There are way too many interactions between nutrients and minerals to continuously adjust for new ingredients. He points out that the amount of potassium eaten in alfalfa-hay would kill cows if given as oral potassium chloride. Veterinarians must be vigilant as new supplements come on the market. They need to make sure that these products are truly working rather than creating damaging chemical chain-reactions.

Today, most of my father's clients try to pay better attention to the amounts of selenium, Vitamin E, and iron in their feedstuffs. Unfortunately, there is nothing you can do to fix this nutrient supplement problem for a baby pig born without the right blood components. For this reason industry experts recommend injecting pregnant sows with a vitamin E/selenium mixture to prevent future problems. Some confinement operators also put a product called Muse, which contains selenium and vitamin E, in the pigs' drinking water.

It seems strange then that mulberry heart disease is becoming a problem for fast-growing hogs as well, because so many measures are being taken to prevent it. Even when additional vitamin E is added to hog feeds, some hogs are dying of mulberry heart. Likely, the oxidative properties of additional feed supplements are destroying the effectiveness of vitamin E and iron in the developing pig. Alternatively, a lack of vitamin D, the sun vitamin, may affect a creature's ability to fully utilize available vitamin E.

As you may recall, vitamin D is a concern in confinement units because many pigs lack adequate amounts of sunlight. In fact, in some confinement units my father can visibly see that hog bone health is better on the south side of a building than on the north side. The pigs on the south side get more sunlight through the curtains. Vitamin D deficiencies could exacerbate problems with other vitamin and mineral regimens because vitamin D affects all of the cells in a pig's body. It also plays a huge role in a pig's immunity to disease.[25] Interestingly, pigs with PFTS are usually pale and anemic-looking, two of the only symptoms reported prior to the development of mulberry heart. We hypothesize that both of these issues could be linked to low vitamin D levels in confinement populations.

In fact, my father has noticed a few improvements in piglets after treating them with a new oral vitamin D product that contains peanut oil.[25] The piglets administered this product tend to do better in general: they have fewer cases of common hog viruses, rickets, and vitamin E deficiencies following weaning.

Despite successes with such products, he has to laugh to keep himself from crying. "It's a mess," he says. "Many of these vitamin D, E, selenium, and iron issues could be prevented if pigs were allowed a few hours outdoors each day to get what they need from the soil and sun, as well as a few extra days to nurse off their mothers to get the beneficial milk they need for a healthy start. Instead, we push them so hard to meet a well defined bottom-line that we can't possibly expect them to be healthy. Then I get to post baby piglets and six month old fat hogs that have died of heart attacks!"

Mycotoxins can also affect a pig's immune response and susceptibility to mulberry heart and PFTS. In addition to destroying the free radicals we associate with cancer and aging, vitamin E destroys some of the toxins and fatty acids associated with poor quality, rancid feeds. When feeds are high in mycotoxins, more vitamin E may be needed to reduce mulberry heart symptoms, including heart tissue lesions. Therefore, chemicals that increase the presence of mycotoxins likely exacerbate issues like mulberry heart and PFTS. These chemicals and mycotoxins weaken the animal's overall immune response.

In one herd my father has treated, mycotoxin (fumonisin) levels were so bad that several large hogs died. This was evident in post-partum analysis because the left-ventricles of the dead hogs' hearts were so muscularized and swollen that only a thimble-sized chamber was left for blood-flow. The hogs' hearts had to overwork to force blood through their systems. When my father measured the ratio of the left-to-right ventricular wall widths, the average was almost three times the ratio in a healthy hog! No amount of medicine was going to fix a heart like this.

So what comes first: the chicken or the egg? Were the mycotoxin levels so high that the pigs' hearts were stressed before other symptoms developed or were the Vitamin E and D levels so low that the pig never had the materials necessary to develop normal sized heart chambers in the first place?

As humans have started to use supplements more frequently, animal farms have had an increasingly difficult time finding and affording quality supplements. Even though they come from the same supply chain, the highest quality supplements go to human markets first. The active nutrients in these products are processed with various chemicals that will increase their storage and transportation lives. Some of the ingredients in these products are bioavailable and some are not.

Most supplements are more expensive than fresh, regional foods our digestive systems have evolved to break down. Some farmers and humans can afford high quality supplements and some cannot. For this reason, I hope that making fresh foods regionally available remains a national priority. We can't continue to treat our pigs and their feed supplies like easy to squeeze piggy banks. If we improve the nutrient content of all foods by respecting and relying more on what the earth naturally gives us, while encouraging the diversification of locally produced fresh goods, the prices of biologically available nutrients will remain lower in the long run.

Unfortunately, we've only touched the surface of nutrient absorption complexities. Certainly we have entered a new era of supplement dependencies ripe with difficult to manage implications: the treadmills taking off beneath our very feet. By improving living conditions and using higher quality supplement sources, we can prevent baby piglet heart attacks before they start.

DIAGNOSIS 10: LAME FEET

For the last three years, my Dad has rarely stopped talking about diseased conditions related to current industrial ag practices. He is obsessed. These are many chronic problems that he knows how to fix but, within current contexts, cannot because doing so would put his clients' businesses in jeopardy. His cures only work in an ideal world where ag policies and practices make common sense.

Obsessions with a particular illness, especially after he's seen *improved* industrial structures wreak havoc on animal health, have changed over the years. When I was ten, I spent many family dinner conversations listening to him gripe about lame cows—cows who for some reason developed bad feet. He often talked about new techniques to treat lame cattle and his desires to help relieve their suffering. He relived dairy herd health conferences, letting the family know what he had learned about laminitis, foot rot, foot and mouth disease, strawberry warts, and other unusual-sounding hoof ailments.

Finally, one day, all of this anxiety filled conversation came to an end. My father drove home with a brand new chute, specifically engineered to help him treat lame feet. He custom designed and built it with the help of a church friend. A man who was sick of bending over to trim so many hoofs in positions that were both awkward for him and the cow was about to see some major relief. I remember thinking that I hadn't seen him so excited in months!

Like a regular chute, Dad's hoof trimming chute has a head gate at the front of an eight-foot long metal platform. The platform rests teeter-totter-style across an axle attached to two wheels. When an animal walks into the back of the chute it must eventually approach the head gate. The head gate is attached to a catch and release lever that a helping hand operates, lest the animal walk straight through and need to be recaptured.

Unlike a regular chute, there are no side panels on Dad's hoof chute. This is so he can reach between side rails to address hoof issues. There is also a soft, wide harness, hooked to a wheel and pulley near the head gate that a second helping hand anchors under the cow's belly. Once the harness is secured, the wheel is used to ratchet the cow up slightly. This discourages the cow from lying down while the veterinarian trims its hoofs.

As I mentioned earlier, cows have very strong legs. Even with this chute, hoof trimming isn't easy. Dad taps one leg to get the cow to lift its hoof up, then uses his favorite lariat rope to anchor the leg to one of the chute's side rails. It then takes him five to ten minutes to work down each deformed or damaged hoof into a more desirable shape with his hoof knife. Still, a second set of hands is needed to hold the rope. If the cow gets upset and kicks, it could easily injure the trimmer, or worse yet, cause the trimmer to gouge himself with his hoof knife.

Dad talked about his new chute for months. It was saving his back. It was helping him reach more feet in need of a trim. It was opening his eyes to whole herds with more foot ailments than he ever wanted to know about.

It was also in high demand. When my Dad obtained his chute, approximately 35% of his clients' dairy cattle were lame, a measure similar to those referenced nationwide.[1,2]

Lameness can be a serious problem in dairy herds because poor hoof quality can result in poor performance and economic losses for dairy farms. By treating feet problems, farmers can increase milk yields and reproductive performance while reducing concurrent issues linked to poor hoof quality like mastitis and abscesses.

Feet ailments can actually be one of the most expensive herd health problems on modern farms, costing anywhere between $70-130 per treatment. In fact, farmers will often cull cattle with reoccurring feet problems because the economic benefits of eliminating lame cattle far exceed the costs of treatment. When a farmer "culls" a dairy cow, he or she essentially removes her from milk production. A culled cow with lame feet usually becomes hamburger.

Lameness can present itself in a variety of ways and have a variety of causes. Nutrition, management and genetics can all predispose a cow to foot problems. Most problems affect the rear feet and present themselves as either an inability to walk properly, deformed hoof claws, warts on the hoof pad, or foot rot.

For the purposes of this discussion I will focus on how proper nutrition management can reduce foot problems. Most cases of lame feet occur within the first 8-10 weeks after a dairy cow has a calf. Dietary stressors that cause lameness can often take such a toll during this time that postpartum cattle develop D.A.s, acidosis, or mastitis. These are all conditions I have helped my father treat on numerous occasions.

Dietary stressors that lead to poor hoof quality can be difficult to manage when the largest percentage of one's diet contains cheap carbohydrates. One of the reasons my father has so much respect for remaining dairy farmers, including those operating successful confinement units, is that these farmers have to be expert nutritionists and workaholics to avoid many of the ailments that would normally affect cows being fed high-carbohydrate, supplement-dependent diets. On the same token, he recognizes that much of this knowledge would not be necessary if the somewhat absurd demands we expect of the modern dairy cow hadn't become expectations in the first place.

To run a successful modern dairy, dairy farmers have to be very familiar with their total mixed rations (TMRs). In a TMR, all animal feed products are mixed together to include forages, concentrates, protein supplements, minerals, and vitamins. By utilizing complete rations, the farm operator can save labor and feed costs because the feed mixture can be delivered all at once. Because it is thoroughly mixed, the farmer can be relatively sure that the cow is obtaining all of the ingredients he or she included in the ration as long as the cow isn't being picky about what she eats. Depending on the point in a cow's milking cycle, the farmer may need to adjust the ration to ensure that the proper balance of dry matter, fiber, calcium, phosphorus, and other vitamins and minerals are delivered to each cow.

Still, the science of ration balancing is tricky. Modern cows didn't evolve to eat the high carbohydrate, protein-rich diets they need to produce the 80-100 pounds of milk we now expect of them each day. To feed carbohydrates correctly, the dairy farmer must pay close attention to the fiber content and particle size of a dairy cow's ration. Feed nutritionists recommend that grains should not exceed 40-45 percent of the overall ration. Ration specialists analyze fiber content values and recommend that fibrous materials, like silage and hay, be chopped so that at least 25% of the particles are two inches long, mimicking larger plant materials cows would eat on pasture.[1] As I mentioned earlier, these larger feed particles increase cud chewing and saliva production. The saliva contains important enzymes that help the cow digest its food.

If the grains aren't properly processed they can upset the rumen and decrease overall performance. Cows expected to calve within three to four weeks need an altered diet to reduce the stress associated with high-grain diets during pregnancy and calving. After the calf is born, the cow still needs time to recover before she is placed back on a high grain diet.

Changes in normal feeding regimens can seriously influence hoof health. When cows are fed rations high in concentrated, non-fibrous materials, especially in their postpartum phases, they are more susceptible to inflammation of the foot, a general condition veterinarians refer to as laminitis. Acidosis predisposes dairy cattle to laminitis because this condition disrupts blood flow to the foot.

Disrupted blood flow to the nerve and vessel-dense area beneath the hard hoof, an area called the hoof corium, reduces overall hoof quality. These conditions also irritate the cow. If the inflammation is severe enough, soft tissues separate from hard tissues setting the stage for more severe infections. The corium can die, hemorrhage, and swell, to the point that a cow can no longer walk. In rare but extreme cases, lame cows and horses are euthanized because they are in so much pain.

Cattle can also develop an irregular gait, which only disfigures the hoof even more. Abnormal wear and tear on different portions of the hoof increase the likelihood that a cow will develop ulcers and abscesses. These conditions can then infect interior ligaments and joints.

Preventing and treating laminitis on modern farms is an on-going battle. However, this is one battle my dad is getting better at fighting. He can often reduce the incidence of laminitis by recommending adjustments to rations and encouraging clients to regularly monitor hoof quality. He encourages these same farmers to provide enough space for cows to easily move around in. Most importantly, he instructs his clients to allow their cows a break from grain intense diets in the days following delivery. Fibrous dry hay and grass are better nutrition for recovery. Still, he wonders if even more dietary changes wouldn't prevent several laminitis issues.

In addition to laminitis, diet can influence the development of foot rot and hairy warts. These conditions are associated with pathogens that enter and affect cows' hoofs. Foot rot is a contagious disease, primarily observed in confinement cattle. It develops when gastrointestinal bacteria are shed in the cow's feces. The bacteria continue living in the fecal matter and surrounding environment before reentering the body through the hoof. Here these bacteria cause swelling and lameness. If the cow is living in a moist environment and the skin between the hoofs gets perforated, there is an additional risk that the cow will contract this disease.

Because the bacteria that cause foot rot typically thrive in low-oxygen conditions, the best way to treat foot rot is to trim and clean the hoof so that more oxygen can reach infected tissues. To kill the bacteria, infected animals are often treated with antibiotics like penicillin. Farmers might also use a disinfectant water bath containing a chemical like copper sulfate to manage problematic bacteria.

Like foot rot, hairy warts are caused by bacteria (though some think that a virus could also be involved). When a cow develops hairy warts, a reddened circular area usually forms on the heel of the rear feet. These warts are raised slightly and have hairs around their edges, hence the name "hairy wart heal rot." Occasionally, this painful condition can also affect the front feet and sole. Cows actually stand on their tiptoes to elevate the infected area and avoid walking when possible. Once diagnosed, treatment needs to begin immediately to reduce the likelihood that the warts spread throughout the herd. Dairy farmers are advised to keep the herd isolated to prevent existing cases from infecting new herd members.

Though foot rot and laminitis can be difficult to treat and manage, hairy warts are the hardest to deal with of the three. While farmers have tried a variety of footbath solutions and antibiotics to cure and reduce the pain associated with hairy warts, research confirming the effectiveness of these treatments remains largely inconclusive. Unfortunately, despite efforts over the last several decades, hairy warts continue to affect most modern dairy farms.

This condition is becoming more common throughout the world, especially in new herds. Less than thirty years ago, the problem was non-existent. It was first observed in Italy before spreading into Europe and, at some point, entering herds in California. From there it spread into herds throughout the nation as cows were sold from one farm to the next. When dairies consolidated in the 80s and 90s, the intermingling of consolidated herds helped spread the disease to more cattle. If one herd was infected, everyone in the newly consolidated herd, which sometimes consisted of ten or more smaller dairies (that had all been bought out) got it. In fact, my dad says that the only clients who don't have hairy wart issues in his practice are those that didn't consolidate and commingle cattle during the last several decades.

In general, there is a positive correlation between herd size and hairy wart prevalence. Studies estimate that 30% or more of the dairy cows sent to American slaughterhouses have hairy warts. This is concerning because warts are the most expensive variety of lameness to treat. They also cost American farmers between 20-50% of the infected cow's milk production.[3]

Because hairy warts are more common in new cattle than in existing cattle, veterinarians suspect that some level of antibiotic resistance may be involved. To treat hairy heel warts, veterinarians use a variety of regimens. As with other types of lameness, trimming is typically required to open the hoof up to oxygen. A variety of topical broad-spectrum antibiotics and medicated footbaths are then used to manage and reduce the spread of infection. In smaller herds, topical antibiotics are applied as powders to each hoof. If a significant portion of a large and concentrated herd is affected, topical powders are not a practical option because they are so high maintenance. With larger herds, most farmers use medicated footbaths that cows walk through or stand in for some period of time. The disadvantage to footbathing is that the baths have to be kept very clean. If the footbaths aren't regularly changed and become filled with cow manure and other materials, they can actually do as much to spread the infection between cattle as they do to reduce it.[2,3]

Effective footbaths may contain a variety of disinfectants. My father has observed success treating hairy warts with tetracycline, lincomycin, formaldehyde, and acidified sodium chlorite solutions. Other studies demonstrate that farmers have had success using zinc chloride, copper sulfate, bleach, and oxytetracycline. The University of Washington Extension reports that there is some evidence that footbaths containing 5% copper sulfate or formaldehyde may be as effective at reducing hairy warts as antibiotics like erythromycin.[4]

Unfortunately several of the aforementioned footbath ingredients are toxic to both the environment and the cow's body.[4] If the cow is exposed to too much copper sulfate, bleach, acid, or formaldehyde, these solutions can kill cells and damage tissues. Some of these solutions are also hard on equipment and farm workers.

In addition to these concerns, non-antibiotic ingredients like zinc and copper can reduce the effectiveness of antibiotics if they are mixed with antibiotic footbath solutions. Farmers need to be familiar with these chemicals and their hazards to apply them in a safe manner that doesn't increase antibiotic resistance.

While Dad has had success managing hairy warts, he admits that his success is largely defined by the short amount of time a modern dairy cow remains productive—a period of typically no more than two years. Because problems continue to present themselves in new herds, he also wonders how effective current treatments will be ten years from now.

In his experience, herds that are still fed some long-stem fibrous plants, like hay and grass, in addition to the ground up particles in their total mixed rations, experience fewer foot rot and hairy wart issues. Similarly, clients who regularly turn their cattle out to pasture, give their cows' feet a break from the cement floors and damp conditions that can exacerbate foot rot and wart problems.

Because bacteria thrive in moist, wet conditions, many well-managed dairies are kept so clean that you could practically eat off the floor. Keeping these dairy facilities clean prevents the bacteria in cow manure and other organic matter from entering the foot through cracks in a cow's hoofs—it's first line of defense against pathogens. Still, many of my father's clients managing well kept facilities, need to use regular footbaths to keep warts under control.

Unfortunately, a clean environment doesn't necessarily aid one's gut. As I discussed previously, the bacteria involved in foot rot and foot warts thrive in low oxygen environments. In addition to moist areas of barns, low oxygen environments include the cow's anaerobic gut.

One of the bacteria associated with hairy warts is called treponema. This bacteria is best known for its symbiotic relationship with a protist that lives in the gut of a termite. This symbiotic protist helps termites digest cellulose in woody plants. It seems possible then that the treponema in a cow's gut also serve some beneficial purpose we don't fully understand. Perhaps, like other bacteria we've discussed, they populate a healthy gut but overgrow intestinal spaces when they are negatively affected by a cow's diet. If a cow, say, weren't getting enough cellulosic fiber, could these bacteria and their symbionts look for food elsewhere and wind up infecting the hoof via manure or the body's own tissues? Perhaps some other aspect of a total mixed ration designed to turn a regular cow into a super-cow upsets the gut's normal bacterial balance?

Not only are cows eating fibrous materials that differ in particle size and content, milk cows today are also eating more protein. My dad has observed that cows eating high concentrations of soluble proteins, which are converted rapidly to free ammonia in the cows' rumens, often have softer padding on the bottoms of their hoof soles. These cows also have less fat under the soles to act as a protective cushioning when they walk. The extra protein is either used to make milk or eliminated but doesn't seem to make its way down to the hoofs in the quantities he has historically observed.

My father thinks that this may have something to do with the way protein levels affect nutrient balances in the blood, causing swelling in the hoof area. When ammonia ions, which are positive, get too high, the body tries to stay electrically neutral by releasing negative hydroxide ions into the bloodstream. This can create problems because the blood becomes alkalotic.[5,6] The body then tries to save positive hydrogen ions to recover from alkalosis while urinating off the most mobile extra-cellular ion: sodium (Na+). This is problematic because sodium works in balance with potassium in many cell transport processes. When potassium builds up inside cells, there is relatively little sodium left outside these cells to counterbalance it. Potassium ends up drawing fluids into the cell causing the cell to swell. This may leave swollen, sodium-deprived regions more susceptible to hoof damage.

Horse experts, see this type of swelling routinely in newborn foals as well as their mothers when they are fed diets high in nitrogenous protein compounds (a condition called founder).[5,6] In fact, two of the most common causes of laminitis in horses are carbohydrate overload and nitrogen-compound overload. In my father's experience, he has seen improvements in hoof quality after convincing dairymen to reduce the soluble protein content in animals' diets. He also encourages these clients to increase the amount of fibrous plant matter in the animals' total mixed rations.

Food safety experts will tell you that milk quality has never been better.[7] That the pathogens in milk are lower than they have ever been. This is probably true. As I've already mentioned, we are pasteurizing more dairy products and sterilizing our barns to kill both the good and bad bacteria that can infect a cow. Unfortunately, this happy little story is only half true.

When the antiseptics and antibiotics used to treat cattle's feet are excreted, they meet other bugs working on a cow's insides, which sometimes re-infect cows through the only surface that protects them from whatever sits on the floor: their hoofs. When cow manure is later dispersed over the soil with other antibiotic pesticides, these bacteria meet even more bacteria in our soils.

Only the most resistant pathogens will survive to recolonize animals. This may occur via two routes: via an open wound or via ingestion of antibiotic-pesticide treated feedstuffs. When there are few good bacteria left, we can't expect a cow, who's already pumped up on high-protein, low cellulose diets, and artificial growth hormones to produce Lance Armstrong-esque dairy miracles without occasionally becoming a victim of these pathogens itself.

This is because a cow's body does not end at its hoofs or skin. It depends on bacteria, protists, and fungi in our soils, manure pits, guts, and facility surfaces. Yet we continue treating the cow's body as an isolated unit promoting resistant pathogens that influence cow health. At the same time, we ignore the likelihood that these pathogens will one day end up repopulating our food supply or other free-living organisms in the environment. Any species that has evolved resistance to antibiotic pesticides, disinfectants, and drugs is a compromise to our food safety system and should be treated as such. These species include plants, bacteria, and animals living both within and outside modern confinement facilities.

My father is most disappointed by "marketing gimmicks" used to promote products and practices that will make cows more efficient and productive, because these gimmicks don't take overall herd health into consideration. From what he can see, some of his middle level-producing herds are the healthiest and require the fewest off-farm inputs including antibiotic and antiseptic treatments. One of the clients he respects the most gets just 75 pounds of milk per day out of his cows, but his cows are noticeably healthier than cows on farms pushed to produce 90-100 pounds per day. His total bottom line is at the top of the pack.

However, the clients getting just 75 pounds per day only remain in business because they ignore industrial efficiency and productivity stats. These clients have beat the system by working hard and making good use of locally available resources, primarily those grown right on their farms. Unfortunately, these practices aren't presented as economically viable and respected options in many farm publications. The industry is busy pushing products to achieve even higher milk productivity stats.

While one of his clients getting 75 pounds of milk per day spends just $4.60/day to feed each cow, other clients getting 85-100 pounds often spend almost twice that to supplement feed rations with off-the-farm feed sources. Statistically, however, these matters go overlooked in itemized dairy reports set up to make individuals believe that industrial dairies are all smooth running, environmentally friendly machines. We then eliminate losses from our stats to make dairy cows appear healthier at these production rates.

When a cow in a super productive dairy isn't performing well, we *cull* it, selling it to the butcher or another farm. Even though this cow's health may have been awful, we don't include it in our annual losses because it didn't die, it's just done working. I've seen all kinds of different cull versus death studies and I'm not quite certain what percentage of our dairy cows die each year versus move on to other farms. My father, who works with cows on a regular basis, will confirm that cull-rate data is ambiguous to say the least. When 20% of a herd needs to be culled at just two years of age because so many animals are crippled with lameness, such statistics should be included. In many cases, they are not. We assume that the cattle were culled for any number of reasons and the public is left in the dark.

Imagine if a school system could cull certain students' standardized test scores. I'm sorry: we already do this. We set up charter and private schools that can more easily reject disadvantaged students. Then we admire their high-performance measures. We forget that many of our public schools don't have the power to weed out populations that may have a more difficult time taking standardized tests. We cull the data and act like remaining public school populations are still representative of the larger population's true abilities.

If it seems that we are becoming more skeptical of researched productivity measures, that's because we should be. There is a reason teachers across the nation are speaking up about statistical injustices. Farmers need to be doing the same. While cows work their hoofs off to feed us, statistical misrepresentations extend even deeper into systems than we've previously imagined.

Collectively, we have overlooked too many things. While we've encouraged the consolidation of herds containing a variety of unique bugs and genetics, we've also encouraged the consolidation of research centers designed to protect us from these very problems. We even have educational programs designed to teach students about the benefits of modern industrial ag—curricular materials made by ag industries themselves. These curricular materials are provided to our nation's financially strapped public schools free of charge so we can teach every child about the benefits of genetically engineered cows and their feedstuffs, while silencing the fact that these benefits might also destroy their feet, put pressure on their guts, and increase antibiotic resistance.

As a nation, we will have to increase our ag research budget and expectations if we hope to address health problems in our plant and animal food supplies. Out of last year's 3.6 trillion dollar budget, almost 20% was allocated to defense spending. The United State's defense budget was more than eight times higher than any other nation and represented approximately 40% of total world military and defense spending. During this same time, the U.S. allotted less than 2% to agricultural spending, most of which was set aside for nutritional assistance programs and crop insurance.[8] Funding allocated for food production research was comparatively miniscule. Over ninety percent of what was allotted was geared toward plant agriculture, mainly corn and bean production. Sustainable agriculture and livestock health represented a tiny sliver of this discretionary pie. You would think that Americans like to eat guns, drones, and seed corn!

Unfortunately, from what I've read in various herd health magazines, much of the food production research that is funded is sub-par. Multinational ag companies have learned how to shape public perception by corrupting science and influencing political decisions.[9] In addition to these well-documented happenstances, important safety research is restricted under current patent protection rights. As I've previously mentioned, special permissions must be granted to independent researchers hoping to compare patented seed varieties to non-patented varieties. If seeds were considered a public good, researchers could better explore how GM feed, antibiotic pesticides, and other chemicals influence laminitis and other diseases, because these seeds wouldn't be subject to legally binding patent limitations.

Research quality is also affected by consolidation in general. Today Monsanto has some control over 90% of GM seed planted globally. Much of this seed is incorporated in animal feed products.[9-12]Those who used to save seed from year to year have struggled to remain competitive with big businesses like Monsanto. In fact, agribusinesses have sued several farmers for saving patented seeds that have contaminated their fields. Mind you, these seeds are so small and numerous that no farmer has the time or funds to sort one cross-contaminated variety from the next. The Public Patent Foundation has filed suit on behalf of 270,000 people from sixty organic and sustainable ag associations asserting that such patents should be unlawful.[9-11]

Yet, American high courts continue to uphold that GM seeds are "substantially equivalent" to their non-GM counterparts and that industry has a right to patent and sell sexually reproducing life forms. These seed monopolies have led to significant increases in corn and soybean seed costs in the last decade.[9-12]

I find it interesting that we have trusted one company with so many research permits to develop these patents in the first place. Between the years 1987 and 2012, the USDA granted Monsanto six times the number of GM research permits granted to its top competitor, Pioneer. In fact, Monsanto was allowed thousands of permits.[13]Since approved, these research endeavors have had far-reaching consequences. When companies like Monsanto and Dupont-Pioneer introduce patented seeds into foreign markets and set up research stations to *help* locals investigate the true potentials of locally adapted seeds, each country needs to clearly establish ownership rights over its seed resources. These generous foreign entities are likely as interested in developing new seed products as they are in helping these nations feed themselves.

Unfortunately, all World Trade Organization (WTO) member-nations are bound to a document in which they agreed to make patents available for technological inventions including seeds used to make animal feeds. According to the WTO's agreement, countries can enact one of three options to protect their plant varieties from foreign research and development. Unfortunately, due to a lack of awareness, ambiguous wording, and the size of the property rights agreement, many nations, particularly developing nations, did not develop effective systems to protect their seeds after the agreement was made. Agbiotech companies were already many steps ahead of these nations in the seed-protection process.[14]These companies have patented genes from around the world and sell seeds wherever they can, wiping out native ecosystems and local businesses along the way.

Even at the national level, patent writers have used ambiguous words to broaden the scope of their patents. Take Monsanto's Roundup Ready herbicide—when the Roundup patent expired in 2000, the company wrote a new patent to maintain control over Roundup Ready technologies.

To do so, Monsanto obtained ownership rights to the practice of premixing glyphosate with other herbicides. This was a practice many farmers already did on their own, because, over time, glyphosate proved less effective against new herbicide resistant weeds. Today, Monsanto's patent covers glyphosate mixtures applied to

> the presently-known glyphosate-tolerant corn, cotton, soybean, wheat, canola, sugarbeet, rice, and lettuce, and any glyphosate-tolerant crop species that may be developed. Also, although the development of glyphosate-tolerant plants by use of conventional breeding without recombinant DNA techniques is currently believed to be highly unlikely, if any such naturally glyphosate-tolerant plants are developed they would fall within the scope of the current invention.[15]

These words landed the company a patent covering a wide range of inventions that didn't even exist yet!

We need more safeguards to make sure that modern ag research, especially research related to highly controversial technologies, is not being spoon-fed to patent agencies, regulators, and the medical health profession by the very individuals benefiting from this research. We would not put the fox in charge of the henhouse then leave for a year and expect to come back to a healthy flock of chickens.

To do so, consumers must demand that the Supreme Court reverse several decisions covering patent policies on sexually reproducing plants (2002 & 1980). Farmers both here and abroad should be allowed to save seeds used for feedstuffs on their farms. Furthermore, researchers should have unlimited access to approved seed varieties that enter public markets. Prior to patent decisions made in 1980, Congress refused to grant patents to seed companies.[16] Until Congress or the courts reverse such decisions, industrial patent holders can sue farmers for either intentionally or *unintentionally* saving seed. Our national patent laws currently do little to protect consumers and farmers. Instead, these laws keep power concentrated in the hands of a wealthy few attempting to control our food supply.

To make such changes possible, we must also do a better job of screening out conflicting interests within our regulatory agencies, industries, and research institutions. In the aforementioned Supreme Court cases, powerful decisions were made by industrial insiders. These decisions never came before voters or congress. In the case deciding modern patent protection, the 2002 Supreme Court decision was written by Justice Clarence Thomas, a former Monsanto attorney. Similarly, the FDA formed regulatory policies for GM seeds under Michael Taylor, a former attorney and public policy director at Monsanto.[16]

Independent researchers hoping to research existing feed supplies, cannot do so if companies do not supply important data on patented products. While investigating total mixed rations, I have learned that most animal feed research is designed to boost productivity. With public support for ag research sitting at such low amounts, independent researchers will have to turn to alternative funding sources to address serious animal and human health concerns.

For these reasons and others, treating lameness is so much more complicated than it needs to be. In addition to normal risk factors, we are predisposing cattle to lameness in ways that deserve much more attention in the research community. Dietary imbalances and antibiotic resistance are among the many risk factors we should evaluate to more effectively reduce these painful and expensive conditions.

DIAGNOSIS 11: HERD HEALTH

We are about halfway done when the rains come. I watch towering thunderheads roll in from the West and tighten my grip on the handle. The cattle shuffle back and forth, knocking each other into the wooden corral, whipping their tails at flies. They know that the storm is coming.

Only the oldest amongst us seems indifferent to the rain. He helps his grandsons drive numbers Sixteen, Forty-two and Eighty down the alley into the chute. I look the front-runner in the eye and let my arm fly up before snapping it down to close the head-gate. I feign confidence but am anxious, worried I might miss the catch or injure the steer. This is my first day on the job and hired workers seem unimpressed that a puny girl has been assigned a position at the chute.

I have been interested in herd health jobs (chutework) since I was six, when my dad came in for career day and discussed herd health with my first grade class. I was somewhat embarrassed at first because he made my entire first grade class act like a herd of cows, but my classmates loved it! He ran us up a make-believe alley and drove us into a cardboard chute before checking us for make-believe illnesses. Some kids wanted to be dehorned, vaccinated, and treated twice.

Today is the real deal. Dad moves quickly. In a matter of seconds,

Sixteen is dehorned, tagged, and vaccinated. Herd health checkups for cattle operations are so routine he could do half of this work with his eyes closed. During this time, I hop on top of a bale alongside the chute to administer one vaccine and the pour-on. Then I scramble down and wait for my cue to open the head-gate. The next steer rumbles forward.

Before we have treated half the herd, a streak of lightning cuts across the horizon. The rain picks up and the cattle push their way to the most protected end of the corral. The bull becomes agitated and a few, tired of fighting the crowd, gallop around the pen's outer edge.

The farmer, a weatherworn character in his sixties, looks worse for the wear. He wants to finish the job but doesn't want any unnecessary injuries. He curses the weather as he opens the back gate to the corral and a river of red and white Simmental cattle flood the pasture. It's dramatic and beautiful. The cattle quickly find shelter beneath an enormous oak at the back corner of the lot.

A farmhand motions toward an old grain wagon near the coral. I help Dad protect his medicine bottles and tools under a couple of metal pails, then run for the wagon. The boys head toward the same end carrying a large, red cooler.

Curious, I watch as one farmhand unhinges the wagon's side panel before, one by one, we crawl in. Other than a small pile of corn and a few buckets, the wagon is empty. Two four-by-six-foot calf panels and a piece of tin lie across the top protecting us from the torrential downpour. A small slit between panels provides a sliver of cloudy-day light. We each grab a bucket, flip it over, and take a seat. Then we open the cooler to find a bag of ham sandwiches, apples, and lemonade—compliments of the house. I rub my wet palms against the cleanest patch of overalls I can find, then dig in before the food disappears.

We eat in silence for several minutes, listening to the roar of rain beat against the tin. It is much too loud to hear anyone anyway, even though some of us are just inches apart.

Eventually, downpours turn to showers and someone cracks a few raunchy jokes. Dad follows these up with one of his infamous rainy day vet stories about the time he examined 52 dead cows on a single vet call. The cows, who had all been standing under a big tree, in tall, wet grass, found themselves in dire straits when a lightning bolt struck the trunk. The cattle, ankles deep in swampy muck, couldn't escape the current that ran down the tree's roots, through the water, and straight up into their legs! The farmer awoke the next morning to find 52 bloated bodies strewn across his field. The only cow left standing was a cripple, a cow he had almost sold the previous week. There was nothing he could do but post a few and try to help the farmer collect insurance.

The farmer muttered a good-humored prayer for his own cattle standing under the oak tree. Dad explained that cattle, who have much more water per body mass and four feet in the muck, are unlikely to be as lucky as the golfer struck down on a golf course. Still the incident seemed especially unlucky.

"Luckily the farmer had enough income to survive the *shock*!" the farmer chuckled. We laughed. The conversation turned to more serious matters as he contemplated his own farm's ability to survive such an incident.

The hired hand later reported that the Swiss Valley Creamery, one of the largest employers in the county, had decided to close. This led to another series of speculations about which farms could weather another change. Many farmers depended on Swiss Valley for additional income.

It has now been over fifteen years since that rainy afternoon but I still think about it. The memory continues to metamorphose as I contemplate complex ag issues. I remember feeling proud of my work that day because I proved that I could keep up with the men. I was also thrilled to be part of the excitement: a massive storm, an unexpected feed wagon luncheon, jokes and vocabulary I'd otherwise never hear....

But the day was also much more than all of this. It was also the first of several experiences in which I drew some connection between farm size and vulnerability. I started to feel the anxiety and stress in clients' sarcastic but honest remarks as we discussed other business closures in the area and how they would impact neighboring farms. On some occasions, concerns were expressed for families of my friends at school—kids I'd played with every day at recess.

I had heard people talk about job loss before but had never really questioned why such changes were happening or who was responsible for them. I just assumed they were inevitable because no one talked very openly about it, especially around kids and adolescents. Not in school or in church. I was too busy thinking about my own dreamed-up opportunities to spend much time contemplating the origins of our rural county's economic woes.

This incredible ignorance of place still frustrates me. While I very much enjoyed assisting my father's veterinary work, I also didn't really understand the consequences of my actions. I didn't initially understand that some aspects of herd-health were much less about health and much more about performance. I did not understand that some farms were conventional by general standards while others operated according to new-fangled prescriptions. My primary concern was to accomplish the task at hand, which I typically assumed, in the case of herd health, was to prevent cows from developing illnesses before they started.

My dad was more in the know. He remembered what herd health jobs were like when he was a kid, and how they'd changed. Clients requested inputs and advice about products that weren't on the market decades before. They needed these items, like hormone implants, to compete even though they didn't necessarily like them. Good or bad, it was the way it was. We assumed our actions were part of an inevitable process.

In some respects, these actions still seem inevitable. The conventional beef farmer, hiring us for this herd health job was not a greedy industrialist or an unfeeling automaton. He was the descendent of a long line of hard working traditional farmers simply trying to make a living in a county where the ability to do so is still largely dictated by external pressures. There were no vibrant farmers markets or CSAs in Delaware County and little demand, at the store level, for fresh, local, and organic foods.

It's ironic: in a county that has historically taken pride in its agriculture, many individuals grow up eating more cheap, imported, and highly processed foods than Americans elsewhere. The county fair and Future Farmers of America (FFA) programs gradually became smaller and smaller throughout my childhood. More kids moved off the farm to increase their chances at attaining less stressful jobs. Today, many of the Fourth of July and Memorial Day parades that used to sport a string of tractors and animals two blocks long don't even have a big enough audience to warrant production.

There is also an abundance of the "this is inevitable" mentality. Farmers who are already locked into enormous debt by investments in machinery, livestock, and buildings can't afford to change. Changing animal ag practices today means much more than risking their personal comfort. Our nation's animal farmers, like any adult citizen, have mouths to feed and children to send to school. Many have had to become very knowledgeable about productivity science and industrial ag economics to survive in this complex system. They can't change in the way that consumers expect them too. They need consumers and politicians to create demand for change first.

While I was browsing through old livestock journals during my last visit home, I came across some advertisements embodying the "this is inevitable" mentality I picked up on as a kid. One advertisement, posted in the mid nineties by Osbourne Automated Systems was entitled "Time to Renovate." In a one page ad, the question, "Time to Renovate?" is centered directly above a golden alarm clock divided into quadrants. In the upper left quadrant is a small picture of the "good old days," various multicolored pigs gathered around an outdoor feeder, one meandering across an open patch of grass in the foreground. The other three quadrants are designated as the prevailing "intense growth era" and contain pictures of uniformly bred hogs confined in Obourne's new and improved automated systems. Below the clock are the words "consumer driven production," followed by "build for your future with Automated Livestock Applications."

Did Osbourne's photos really convince farmers that the "intense growth era" was more promising than the "good old days?" The photo seems to imply a series of rhetorical questions: Are bars better than freedom? Monotony better than diversity? Ugliness better than beauty? And yet, even as the alarm sounds these questions, the ad confidently claims its effectiveness. It whips out the ultimate persuasion: our own gloomy fear that the intense growth era is already here! Even if we do appreciate the good old days more than the era of confinements and monocultures, the ad begs us to assume that we are alone. That in order to keep up with the times and make money we must limit our decisions to economic cost-benefit ratios and forget about less important values like animal health, ecological diversity, and work-ethic.

I am reminded of several articles I've recently read regarding the benefits of GM crops. These articles assure me that consumers are coming to accept GM crops in their food supplies, the same GM crops many of us don't even understand are there! Of course, the era is already upon us and we can either accept it or spend a large amount of time and effort fighting it. The sponsors of such publications hope we'll forget the assumptions made to arrive at this point in the first place!

The simple truth is that many Americans are not equipped to challenge widely advertised assumptions. As a child who grew up in rural Iowa, I heard little regarding alternatives to this advertised version of *progress*. It wasn't until my sophomore year of college that I began to think more critically about the state of American ag after reading a book called Farm as Natural Habitat by Dana Jackson.

The case studies Jackson used to describe relationships between soil, food, and the environment hit home and I began to investigate ways in which farming and conservation might be viewed as more compatible rather than antagonistic. Wanting to explore these views, I interned for the USDA, job shadowed a land-grant soil scientist, house-sat for CSA farms, and worked at Seed Savers, a fruit and vegetable seed bank in Decorah, Iowa. Gradually, I came to appreciate that food is so much more than calories. It is medicine. It is the root of many cultural practices. It is part of the ecosystems that surround it. I also began to question many modern production practices, including those used in my own county.

Though I have spent years contemplating rainy day feed wagon experiences, I typically arrive at the same conclusion: agriculture is complicated and *progress* seems more inevitable for some than others. Most importantly: our prescriptions, whether they be medical or political, *do* matter. The external forces shaping laws and regulations *do* trickle down to regional economies and ecosystems. These forces have varying effects on our nations' meat, grain, dairy, and poultry producers.

My dad likes to remind his clients that prevention is the best medicine. By addressing problems before they start, you can save yourself time, anxiety, and money. This is why animals and humans have routine physicals and herd health check-ups. By catching problems before they start, we can treat illnesses before creatures come into contact with them. Having open conversations, in which we assess *all* of the ways current ag practices affect our safety and health, can help prevent problems as well.

Life is so precious. It is also incredibly interdependent. There is no magic silver bullet that will cure any of the ailments described in this book. But there are many little things each one of us can do to reduce the number of unhealthy humans and animals living on our planet. There are also ways we can influence decisions at federal and state regulatory levels to better address national herd health issues.

In a sense, our laws and regulations are a form of preventative herd health for every creature and citizen. When ag laws and regulations are functioning properly, illnesses can be prevented, along with the wide range of problems that accompany them. When our laws and regulations are dysfunctional, they negatively affect farms and sound the alarm that *progress* is not what it seems.

While my dad knows that many current animal farm practices are not positively affecting animal health, he gets defensive because many of our national herd health laws have done little to encourage health on animal farms. Simultaneously, these laws have encouraged crop practices that have reduced farmers' abilities to feed their animals nutritious diets. We don't even regulate herbicidal antibiotics that are used thousands of times more than animal antibiotics! In fact, we continue to encourage farmers to feed their cattle and pigs cheaply produced, toxin-coated seeds even though we have the capacity to feed them more diverse, nutritious foods. We need policies that encourage farmers to promote health by growing a wider variety of less toxic feedstuffs.

Citizens from around the world, including many Americans, are already demonstrating desires to have more ecological, regionalized ag practices to reduce the health and environmental costs I have come to associate with crop monocultures and factory farms. These people are growing gardens, shopping locally, using integrated pest management strategies, pressuring restaurants to phase out confinement cages, supporting food labeling bills, and preserving seed varieties. They are also creating demand for farmers markets and local, value-added foods in groceries. Sometimes, consumers are even supporting groups like Heifer International to help feed the world instead of investing in corporate land grabs. While each of these actions is significant, we need more political and regulatory changes to discourage monoculture crops and diets dependent on unnecessary toxins.

Some of our ag trading partners are setting much higher expectations. In the Netherlands, for example, consumers are demanding more transparent labels tied to on-farm environmental stewardship and welfare. In addition to crop insurances tied to on-farm diversity, animal product labels give consumers some sense of the animal's quality of life. Animals with better living conditions and less toxin-dependent feed supplies earn a higher ranking in these systems and are priced accordingly. Other countries have banned GM crops altogether and are regulating pesticide use more strictly. Decisions like these reflect an understanding that farm diversification can help citizens better manage pests and pollution in addition to reducing long-term health risks associated with such practices.

Decisions to improve national herd health are not nearly as complicated as the numerous diseases confinement buildings and monocultures have created. In fact, some solutions to my father's herd health problems seem almost comically simple: more cows need to eat grass, more pigs need to get a few hours of sunlight, and we need to stop feeding animals toxins known to destroy healthy bacteria.

On another level, change will take a political pig-headedness many Americans haven't demonstrated since the Great Depression. Ultimately, we have to stop discussing ag problems in terms of economic efficiencies and productivities. When all life clings to some remaining fragment of an ecosystem, life that tries to exist in an area intimately intertwined with millions of acres of uniformity around it, we create unnecessarily complex herd health problems. All of this fragmented sameness hardly begets sustainable progress. Unless we assign diversity and health greater economic values, we will never sustain the soils that have fed us in the past.

We typically teach our kids to be kind, compassionate human beings—good stewards of our future communities and lands—because we want them to reap the benefits provided by their future communities and ecosystems. To do so, we will have to be better, more aware and participatory stewards of the ag-ecosystems around us. Citizens from all social spheres need to work together to urge government leaders to restore food policies that promote crop diversity, jobs, and healthier consumer and producer options at regional levels. It's time, too, that elected officials put their feet down when multinational conglomerates put all creatures at risk. We need to work harder to free ourselves from integrated powerhouses that monopolize marketed values and gene pools that took billions of years to co-evolve!

Unfortunately global conglomerates continue to direct our thoughts and actions toward systems that never made sense in the first place. To make matters worse, we have standardized care practices that coach us on how to change our wording rather than our work! As Upton Sinclair once wrote, "It's difficult to get a man to understand something when his job depends on not understanding it."[1] Many politicians, researchers, scientists, farmers *and consumers* have learned to depend on nonsensical food practices. The result is that we often end up sacrificing health and animal welfare without even knowing it.

We can address these issues, in part, by electing leaders who pay attention to the common good. Not leaders who use words like pawns, one head talking *at* another. Leaders who are willing to question practice, initiate open debate, and encourage day-to-day thoughtfulness. Such leaders understand that we are part of the ecosystems and economies we shape, benefit, and destroy, and that our quality of life depends on each interconnected part.

As other food advocates have pointed out, we have learned how to dissociate what we spend from the farmers and citizens our food dollars affect. In doing so, we can avoid thinking about how our actions affect actual creatures. Such disassociations are most possible in a non-localized world where we can't see the direct consequences of our actions. As a teacher, I've observed this first hand. While my students have the ability to understand how global issues can be good or bad, we typically only have the capacity to really love and care for what is closest to us. Some of us, including seemingly intelligent political leaders and parents, are so distracted by global financial solutions, personal economic opportunities, and technological possibilities that we forget to take care of what's at home: our regions, our bodies, our schools, our children, and our communities. Ironically, these are the aspects of life we can most easily influence.

I suspect that one day future generations will remember the last three decades as a ridiculous age in American agriculture. This has been an age during which too many human beings treated animals and children like guinea pigs, feeding them genetically modified, chemically coated, antibiotic resistant experiments, despite the overwhelming evidence that these foods are serious risk factors for illness and disease. In today's world of widely accessible research and technological advances, the ability to produce abundant amounts of food without threatening biodiversity and our basic biological rights should be an expectation, not a goal.

Our Biorights:

Implicit in our rights to life, liberty, and the pursuit of happiness is our right to optimal health. This means that we:

1. Have a right to a genetic code free of gene pollution—a genome that can coevolve with the world around it, unaltered by foreign, potentially damaging genetic fragments in our food supply.

2. Have the right to know and research what is in our food including genetically modified and toxic substances that may have been used to produce a food product.

3. Have the right to an environment that contains minimal synthetic toxins, especially those toxins that increase one's chances of developing diseased conditions.

4. Have the right to manage our own reproductive health free of corporate products that reduce our sperm counts and diminish the quality of an egg, as well as the DNA passed from one generation to the next.

5. Have the right to a life free from government-supported food discrimination. Unhealthy processed foods should not be artificially cheaper than the raw products used in their making.

6. Have the right to access affordable seeds so we can continue to grow food that truly nourishes future generations.

Co-Author's Notes:

Science that does not ask the right questions is not science. Industry funded science is often designed to research products for sale. Next to no one is educated to look at the whole pie anymore. Everyone is encouraged to specialize. We have to reform education to recruit more general practitioners who understand the interconnections between biological systems and the regulations we use to monitor them.

When I go to veterinary meetings and scientists argue that you have to have statistical p-values to support every diagnosis and treatment I often want to walk out. As a clinician, I understand that you have to watch how you interpret what you see, but if what you see is consistently different from the "science" you are presented with at these meetings, you begin to question this "science." As a clinician, I learn more from my mistakes than from my successes. This is one reason why multiple ownership with numerous small to medium-sized farms is more reliant from a medical perspective. When a health specialist only sees one to two huge set-ups there are fewer variables and mistakes to learn from. If you are a general practitioner working with different operators, different genetics, and different nutrition programs, and you see the same problems over and over again, you can bet that there is more to a problem than meets the eye.

With less state and land grant funding for our public ag research universities, more colleges are using meta-analysis studies and calling them "science." In a meta-analysis an academic looks at many previous studies and sees if over 50% of the research material supports his or her hypothesis. In my book, this is not science at all. I'm typically unimpressed because I know that most of this "science" being meta-analyzed was paid for by industry to promote its use.

- Art Dunham, DVM

SOURCES

Diagnosis 1: American Agriculture's Two-Headed Pig

1. "Agricultural Data For Decision Makers: Delaware County." Iowa State University Extension. 2001. http://www.nass.usda.gov/Statistics_by_State/Ag_Overview/AgOverview_IA.pdf
2. http://www.agcensus.usda.gov/Publications/2007/Full_Report/Volume_1,_Chapter_2_County_Level/Iowa/
3. Vance, Andy. "Vet Medicine Faces Challenges." *FeedStuffs: The Weekly Newspaper for Agribusiness.* June 11, 2012.
4. http://www.epa.gov/agriculture/ag101/demographics.html, Viewed: March 3, 2013.
5. Fatka, Jacqui. "Calls for FRS reform increase." *Feedstuffs: The Weekly Newspaper for Agribusiness.* December 3, 2012.
6. Vance, Andy. "Food prices used in RFS plea." *Feedstuffs: The Weekly Newspaper for Agribusiness.* August 6, 2012.
7. Smith, Rod. "Farmers Get Little of the Food Dollar." *Feedstuffs: The Weekly Newspaper for Agribusiness.* December 31, 2012.
8. EWP Report: Viewed August 8, 2012. http://static.ewg.org/reports/2012/farm_bill/babcock_free_crop_insurance.pdf
9. The Innovative Ag Services Website: http://www.ias.coop/index.cfm Viewed July 4, 2012.
10. http://www.ars.usda.gov/research/publications/publications.html Viewed: March 20, 2013.
11. http://www.bigpictureagriculture.com/2012/07/hot-5-midwest-drought-2012-corn-usage-pie-charts-vilsack-comic-cc-precipitation-maps-faraday-porteur.html, Viewed March 3, 2013.
12. http://www.rodaleinstitute.org/fst30years, Viewed June 2012.
13. http://www.leopold.iastate.edu/pubs-and-papers/2011-11-ltar-experiment, Viewed August, 2012.
14. http://www.cdc.gov/Features/WeightoftheNation/, Viewed January 2013.
15. Smith, Rod. "Producer support of beef checkoff strong." *Feedstuffs: The Weekly Newspaper for Agribusiness.* February 11, 2013.
16. Becker, Geoffery. "Federal Farm Promotion ("Check-off") Programs." *CRS Report for Congress.* October 2008.
17. Wegner, Michael. "Lost in Translation." *Pork Checkoff Report.* National Pork Board. Summer 2012: Vol 31, No. 2.
18. Novak, Chris. "In Defense of Choice." *Pork Checkoff Report.* National Pork Board. Summer 2012: Vol 31, No. 2.
19. http://www.humanesociety.org/assets/pdfs/farm/HSUS-Report-on-Gestation-Crates-for-Pregnant-Sows.pdf Viewed July 2012.
20. Welshans, Karissa. "Researchers Compare Calf Housing Types." *Feedstuffs: The Weekly Newspaper for Agribusiness.* July 9, 2012.
21. Pagan, Joe. "Past, future advances in equine nutrition." *Feedstuffs: The Weekly Newspaper for Agribusiness.* October 31, 2012.

22. http://en.wikipedia.org/wiki/Veterinarian's_Oath Viewed February 14, 2013.
23. https://www.avma.org/KB/Policies/Pages/Principles-of-Veterinary-Medical-Ethics-of-the-AVMA.aspx Viewed February 14, 2013.
24. King, Mike. "Pork's Sustainable Future: data reveal decades of continuous improvement." *Pork Checkoff Report*. National Pork Board. Summer 2012: Vol 31, No. 2.
25. http://www.agcensus.usda.gov/ Viewed March 13, 2013.

Diagnosis 2: Manganese Deficiency

1. Blezinger, Stephen, Phd. 1998-2005. *Obtaining an Accurate Mineral Status Can be Complicated*. Cattle Online Today. Viewed July 8, 2012. http://www.cattletoday.com/archive/2006/April/CT460.shtml
2. Schefers, Jeremy, DVM. *Fetal and Perinatal Mortalities Associated with Manganese Deficiency*. Minnesota Dairy Health Conference. 2011.
3. Funke et. al. August 17, 2006. *Molecular Basis for the Herbicide Resistance of Roundup Ready Crops*. PNAS August 29, 2006 vol. 103 no. 35.
4. The Center For Food Safety (primary author: Bill Freese). Sept. 21, 2009. RE: Registration Review: Glyphosate Docket. Docket Number: EPA-HQ-OPP-2009-0361.
5. USDA: ERS: Recent Trends in GE Adoption. http://www.ers.usda.gov/data-products/adoption-of-genetically-engineered-crops-in-the-us/recent-trends-in-ge-adoption.aspx Viewed August 20, 2012.
6. Pesticide Industry Sales and Usage Report: 2006 & 2007 Market Estimates, EPA.
7. U.S. EPA ReRegistration Active Ingredient List. 5/13/09. EPA-HQ-OPP-2009-0361-0004.
8. The Center for Food Safety: The True Food Network. Viewed 7/10/12. http://truefoodnow.org/campaigns/genetically-engineered-foods/.
9. András Székács and Béla Darvas. Forty Years with Glyphosate, Herbicides - Properties, Synthesis and Control of Weeds, Dr. Mohammed Naglb Hasaneen (Ed.), ISBN: 978-953-307-803-8, InTech. 2012. http://www.intechopen.com/books/herbicides-properties-synthesis-and-control-of-weeds/forty-years-with-glyphosate
10. The Agronomic Benefits of Glyphosate in Europe. Monsanto Inc. 2011.
11. Colburn, Dumanoski, & Myers. Our Stolen Future. Penguin Books U.S.A, Inc, 1996.
12. Monaco, Veller, Ashton. 2002. Weed Science: Principles & Practices. John Wiley & Sons, Pg 326-334.
13. http://www.usgs.gov/newsroom/article.asp?ID=2909, Viewed March 18, 2012.
14. Effects of Roundup(®) and Glyphosate on Three Food Microorganisms: Geotrichum candidum, Lactococcus lactis subsp. cremoris and Lactobacillus delbrueckii subsp. bulgaricus. Curr Microbiol. Feb 24, 2012.
15. Bott et. al Glyphosate-induced impairment of plant growth and micronutrient status in glyphosate-resistant soybean. Plant Soil. 2008, 312, 185-194.

16. Yamada et. al. Glyphosate interactions with physiology, nutrition, and diseases of plants: Threat to agricultural sustainability. Eur. J. Agron. 2009, 31, 111-113.
17. Zobiole et al. Nutrient accumulation and photosynthesis in glyphosate-resistant soybeans is reduced under glyphosate use. J. Plant Nutr. 2012, 33, 1860-1873.
18. Zobiole et al. Effect of glyphosate on symbiotic N2 fixation and nickel concentration in glyphosate resistant soybean. J. Agric. Food Chem. 2010, 58, 4517-4522.
19. Zobiole et al. Glyphosate reduces shoot concentrations of mineral nutrients in glyphosate-resistant soybeans. Plant Soil 2010, 328, 57-69.
20. Zobiole et al. Glyphosate affects micro-organisms in rhizospheres of glyphosate-resistant soybeans. J. Appl. Microbiol. 2011, 110, 118-127.
21. Zobiole, L. H. S.; Kremer, R.J.; Oliviera, R. S., Jr.; Constatntin, J. Glyphosate affects chlorophyll, nodulation and nutrient accumulation of "second generation" glyphosate-resistan soybean Pestic. Biochem. Phyisol. 2011, 99, 53-69.
22. Zobiole, L. H. S.; Oliviera, R. S., Jr.; Constantin, J.; Kremer, R. J.; Biffe, D. F. Amino acid application can be an alternative to prevent glyphosate injury in glyphosate-resistant soybeans. J. Plt Nutr. 2012, 35, 268-287.
23. Bellaloui, N.; Reddy, K. N.; Zablotowics, R.M.; Abbas, H. K.; Abel, C. A. Effects of glyphosate application on seed iron and root ferric (III) reductase in soybean cultivars. J. Agric. Food Chem. 2009, 57, 9569-9574.
24. Means, N. E.; Kremer, R. J. Influence of soil moisture on root colonization of glyphosate-treated soybean by Fusaium species. Commun, Soil Sci, Plant Anal. 2007, 38, 1713-1720.
25. Duke et. al. Glyphosate Effects on Plant Mineral Nutrition, Crop Rhizospher Microbiota, and Plant Disease in Glyphosate-Resistant Crops. J. Agric. Food Chem. 2012.
26. Huber, Don. "What's New in Ag Chemical and Crop Nutrient Interactions." Fluid Journal: Official Journal of the Fluid Fertilizer Foundation. 2010. Volume 18 No 3 Issue 69.
27. Estabrook, Barry. Tomatoland. Pg x. Andrews McMeel Publishing, 2012.
28. Seralini, G; Letter to the Editor. http://dx.doi.org/10.1016/j.fct.2012.11.007 Food and Chemical Toxicology. November 2012.

Diagnosis 3: Post Weaning Failure to Thrive Syndrome

1. Huang Y, Henry S, Friendship R, et al. Clinical presentation, case definition, and diagnostic guidelines for porcine periweaning failure to thrive syndrome. *J Swine Health Prod.* 2011;19(6):340–344.
2. Vansickle, Joe. "Researchers Scramble to Solve Failure to Thrive Syndrome. National Hog Farmer Magazine." Sept. 15, 2008.
3. Huang Y, Gauvreau H, Harding J, Diagnostic investigation of porcine periweaning failure-to-thrive syndrome: lack of compelling evidence linking to common porcine pathogens, J Vet Diagn Invest. 2012 Jan;24(1):96-106. Epub 2011 Dec 6.
4. Moeser AJ, Borst LB, Overman BL, Pittman JS. Defects in small intestinal epithelial barrier function and morphology associated with periweaning failure to thrive syndrome (PFTS) in swine, Res Vet Sci. 2012 Oct;93(2): 975-82. Epub 2012 Jan 28.
5. Sawyer, John. ISU Extension Power Point Presentation. 2008.
6. Dunham, Art, DVM. Interview. July 8, 2012.
7. Johal et al. Glyphosate's Effects on Diseases of Plants. European Journal of Agronomy 31: 2009, 144-152.
8. http://earthopensource.org/index.php/reports/58, Viewed July 24, 2012
9. Animal Health and Veterinary Laboratories Agency. "Periweaning Failure to Thrive Syndrome: Information for Farmers and Vets in Great Britain." Viewed July 8, 2012. https://docs.google.com/viewer?a=v&q=cache:wKIsMEQSTQoJ:vla.defra.gov.uk/science/docs/sci_pfts.pdf+post+weaning+failur+to+thrive+pigs&hl=en&gl=us&pid=bl&srcid=ADGEESjDgg2Os2lxaG7PdVhlm4c-WWeSk3KDk_IL61mvo5hkWMZirOToKaEVflilF51javwSDb7hN47ebzcP-NDkaL9y6Z4
10. Hymen, Mark. The UltraMind Solution. Schribner, 2009.
11. Eker, S., Ozturk, L., Yazici, A., Erenoglu, B., Roemheld, V., Çakmak, I., 2006. Foliar- applied glyphosate substantially reduced uptake and transport of iron and manganese in sunflower (Helianthus annuus L.) plants. J. Agric. Food Chem. 54, 10019–10025.
12. Cakmek et al. Glyphosate reduced seed and leaf concentrations of calcium, manganese, magnesium, and iron in non-glyphosate resistant soybean. Sanbanci University Turkey. 2009.

Diagnosis 4: Infertility

1. http://www.ewg.org/reports/bodyburden2/execsumm.php, Viewed August, 2012.
2. Hyman, Mark. Pg 221 The UltraMind Solution. Scrhibner. 2005.
3. A. Jenks Britt, D.V.M and Fernando Alvarez, M.V.Z. Why are so many cows losing pregnancies. Hoard's Dairyman Nov. 2011.

4. "Testimony of the Ranchers-Cattlemen Action Legal Fund, United Stockgrowers of America, to the Senate Agriculture Committee." July 24, 2002.
5. Ishmael, Wes. "Nowhere Fast: Cow Productivity remains static despite increasing breed genetic trends." Beef Magazine, pg 40-42. Feb 2012.
6. Antoniou, M. Robinson, C. and Fagan. J. "GMO Myths and Truths." http://earthopensource.org/files/pdfs/GMO_Myths_and_Truths/GMO_Myths_and_Truths_1.3.pdf
7. Smith, Jeffrey. Seeds of Deception. Yes! Books, 2003. http://www.i-sis.org.uk/US_drought_destroys_GM_Crops.php
8. Sirinathsinghji, E. GM crops destroyed by US drought but non-GM varieties flourish.: Non-GM varieties more drought resistant, yet agritech giants ensure farmers are unable to access them. ISIS Report. 2012.
9. Latham, J and Wilson, A. Is the Hidden Viral Gene Safe? GMO Regulators Fail to Convince. Independent Science News: Biotechnology: Februrary 27, 2013.
10. "Decline of Livestock Genetic Diversity Being Addressed." *Feedstuffs*. October 29, 2012.
11. Lunden, Tim. "Consider Options for Feeding Challenges." *Feedstuffs*. November 12, 2012.
12. Smith, Jeffrey. Genetic Roulette. Chelsea Green Publishing, May 2007.
13. www.responsibletechnology.org. Viewed: July 8, 2012.
14. http://naturalsociety.com/monsantos-infertility-linked-roundup-found-in-all-urine-samples-tested/. Viewed: August 7, 2012.
15. Toxicol In Vitro. 2011 Dec 19. Epub 2011 Dec 19. PMID: 22200534
16. Arden, Anderson. Powerpoint Table: Foods, Diet, and Nutrition: What a Clinician Needs to Know about Nutrient Density, Pesticides and Genetically Engineered Foods. Presentation on Age Management Medicine, 2011.
17. Smith, Jeffrey. 2010. Genetically Modified Soy Linked to Sterility, Infant Mortality in Hamsters. http://www.huffingtonpost.com/jeffrey-smith/genetically-modified-soy_b_544575.html
18. Ermakova, Irina. Oct 2005. Preliminary Study: Influence of Genetically Modified-SOYA on the Birth-Weight and Survival of Rat Pups. Institute of Higher Nervous Activity and Neurophysiology Russian Academy of Sciences (RAS).
19. Brown, Patrick. "The Promise of Plant Biotechnology." University of California, Davis.
20. Brumfiel, Geoffrey. GM Canola escapes Corn Fields. NPR Morning Edition. August 6, 2010.
21. Latham J. and Wilson A. *The Great DNA Data Deficit: Are Genes for Disease a Mirage.* Independent Science News. Dec. 8, 2010.
22. Biemont, Christina. A Brief History of the Status of Transposable Elements: From Junk DNA to Major Players in Evolution Genetics December 2010 186:1085-1093; doi:10.1534/genetics.110.124180
23. Crews D. et. al. *Epigentic Transgenerational Inheritance of Altered Stress Responses. Proceedings of the National Academy of Science.* May 21, 2012.
24. Sebesta, Steve. "98 Reasons not to use glyphosate on seed fields: Don't use glyphosate as a harvest aid in seed fields!" Dec 6, 2011.

http://www.farmandranchguide.com/feature/seed_guide/reasons-not-to-use-glyphosate-on-seeds

25. Glyphosate-based herbicides are toxic and endocrine disruptors in human cell lines. Gasnier C, Dumont C, Benachour N, Clair E, Chagnon MC, Séralini GE. Toxicology, 2009 Jun 17.
26. Glyphosate formulations induce apoptosis and necrosis in human umbilical, embryonic, and placental cells. Benachour N and Seralini, GE. Chem Res Toxicol, 22: 97-105, 2009.
27. Time- and dose-dependent effects of roundup on human embryonic and placental cells. Benachour N, Sipahutar H, Moslemi S, Gasnier C, Travert C, Séralini GE. Arch Environ Contam Toxicol, 53: 126-133, 2007.Control human embryonic 293 cellsSame cells treated only 24 h with 0.05% Roundup
28. "Glyphosate-based herbicides produce teratogenic effects on vertebrates by impairing retinoic acid signaling" Paganelli A, Gnazzo V, Acosta H, López SL, Carrasco AE. (2010) Chem Res Toxicol. 23: 1586-1595.
29. Glyphosate and Roundup cause birth defects in frog and chicken embryos, in doses far lower than those used in agricultural spraying.
30. http://naturalsociety.com/monsantos-infertility-linked-roundup-found-in-all-urine-samples-tested/. Viewed August 7, 2012.
31. Clair E, Mesnage R, Travert C, Séralini GÉ.Toxicology In Vitro. 2011 Dec 19. Epub 2011 Dec 19. PMID: 22200534
32. Samsel, A, Seneff, S. Glyphosate's Suppression of Cytochrome P450 Enzymes and Amino Acid Biosynthesis by the Gut Microbiome: Pathways to Modern Diseases. Entropy 2013, 15, 1416-1463; doi:10.3390/e15041416.
33. http://www.wildlifeextra.com/go/news/neonicotinoids-birds.html#cr Viewed March 22, 2013.
34. http://www.reduas.fcm.unc.edu.ar/wp-content/uploads/downloads/2011/10/INGLES-Report-from-the-1st-National-Meeting-Of-Physicians-In-The-Crop-Sprayed-Towns.pdf
35. Glyphosate-based herbicides are toxic and endocrine disruptors in human cell lines. Gasnier C, Dumont C, Benachour N, Clair E, Chagnon MC, Séralini GE. Toxicology, 2009 Jun 17.
36. Toshinobu Miyamoto, Akira Tsujimura, Yasushi Miyagawa, Eitetsu Koh, Mikio Namiki, and Kazuo Sengoku, "Male Infertility and Its Causes In Human," *Advances in Urology*, vol. 2012, Article ID 384520, 7 pages, 2012. doi:10.1155/2012/384520
37. Alejandro Oliva, Alfred Spira, and Luc Multigner. Contribution of environmental factors to the risk of male infertility. Hum. Reprod. (2001) 16(8): 1768-1776 doi:10.1093/humrep/16.8.1768
38. Sirinathsinghji, Eva. "Bt Resistant Rootworm Spreads."
39. Benbrook, Charles. Glyphosate Resistance is slipping and unstable transgene expression erodes plant defenses and yields. May 2001, AgInfoNet, Techinical Paper.
40. Carman, J, Heinemann, J, Agapito-Tenpen, S. "A briefing document for non-specialists describing the lack of regulation of a new class of products and GM crops based on dsRNA technology." March 21, 2013.
41. Heidinger, Kurt. *Trasgneic DNA from GMOs in Chinese Rivers - Why is it Suddenly There?* BioCitizen, March 25, 2013.

Diagnosis 5: Schistosomus Reflexus

1. Glyphosate induced cell death through apoptotic and autophagic mechanisms. Chinese Medical School & Departments of Neurology, Health Science, 2012.
2. ISIS Press Release. "Scientists Confirm Potent Hormone Disrupting Effects." 7/15/09. Viewed July 10, 2012.
3. Mason, Rosemary. Letter to Commissioners Dalli, Ciolos and Potocnik and Vice President Ashton. Viewed July 16, 2012.
4. Phend, Crystal. *Pest Poison Tied to Brain Defects in Young Kids.* MedPage Today. May 1, 2012.
5. Pan-Montojo, Francisco; Anichtchik, Oleg; Dening, Yanina; Knels, Lilla; Pursche, Stefan; Jung, Roland; Jackson, Sandra; Gille, Gabriele et al. 2010. Kleinschnitz, Christoph. ed. "Progression of Parkinson's Disease Pathology Is Reproduced by Intragastric Administration of Rotenone in Mice". *PLoS ONE* 5 (1): e8762. doi:10.1371/journal.pone.0008762. PMC 2808242. PMID 20098733.
6. Tanner, Caroline M. et al. (2011). "Rotenone, Paraquat and Parkinson's Disease". *Environmental Health Perspectives* 119 (6): 866–72. doi:10.1289/ehp.1002839. ISSN 0091-6765. PMC 3114824. PMID 21269927.
7. Wallheimer, Brian. "Researchers: Honey Bee Deaths Linked to Seed Insecticide Exposure." January 11, 2012. http://www.purdue.edu/newsroom/research/2012/120111KrupkeBees.html
8. Pilatic, Heather. "GE Corn & Sick Honey Bees – What's the Link." Pesticide Action Network. http://www.panna.org. 4/19/2012.
9. http://www.epa.gov/opp00001/regulating/registering/. Viewed: July 7, 2012.
10. U.S. EPA. Glyphosate Final Work Plan. December 2009.
11. U.S. EPA ReRegistration Decision Fact Sheet for Glyphosate (EPA-738-F-93-011) 1993.
12. The Center For Food Safety (primary author: Bill Freese). Sept. 21, 2009. RE: Registration Review: Glyphosate Docket. Docket Number: EPA-HQ-OPP-2009-0361.
13. U.S. EPA ReRegistration Active Ingredient List. 5/13/09. EPA-HQ-OPP-2009-0361-0004.
14. Beyond Pesticides. Sept 21, 2009. Response to EPA Re-registration Review. Docket No. EPA-HQ-0PP-2009-0361.
15. Adam A, Marzuki A, Abdul Rahman H and Abdul Aziz M, The oral and intratracheal toxicities of Roundup and its components to rats. Veterinary and Human Toxicology, 39(3): 147-151, 1997.
16. Mesnage R., Bernay B., Séralini G-E. (2013, in press). Ethoxylated adjuvants of glyphosate-based herbicides are active principles of human cell toxicity. Toxicology http://dx.doi.org/10.1016/j.tox.2012.09.006
17. Séralini G. E., et al. (2012). Long term toxicity of a Roundup herbicide and a Roundup-tolerant genetically modified maize. Food and Chemical Toxicology 50 (11): 4221-4231. And (3) Séralini G. E., et al. 2013. Answers to critics: Why there is a long term toxicity due to NK603 Roundup-tolerant genetically

modified maize and to a Roundup herbicide. Food and Chemical Toxicology.

18. Tsui MT and Chu LM, Aquatic toxicity of glyphosate-based formulations: comparison between different organisms and the effects of environmental factors, Chemosphere, 52(7): 1189-1197, 2003.
19. Huber, Don. "Direct Toxicity of Glyphosate." Powerpoint Presentation. Viewed July 12, 2012.
20. Gasnier, C., Dumont, C., Benachour, N., Clair, E., Chagnon, M-C., and Seralini, G-E. 2009. Glyphosate-based herbicides are toxic and endocrine disruptors in human cell lines. Toxicology 262: 184-191.
21. Watts, Meriel. "Glyphosate." PANAP: November, 2009.
22. Thicke, Francis. Email to Practical Farmers of Iowa. Viewed December 2012.
23. Pesticide Industry Sales and Usage Report: 2006 & 2007 Market Estimates, EPA.
24. http://www.epa.gov/pesticides/factsheets/riskassess.htm Viewed July, 2012.
25. Beyond Pesticides. Response to EPA Re-registration Review. Docket No. EPA-HQ-OPP-2009-0361. Sept 21, 2009.
26. http://www.aphis.usda.gov/biotechnology/about_mission.shtml Viewed July 2012.
27. Bell & Cole. "USDA Announces Availability of Biotechnology Regulatory Petitions and Other Actions: *Agency Provides Additional Opportunity for Public Review and Comment as Part of Strong, Consistent Regulatory Review Process. USDA News Release.* Viewed July2012. http://www.aphis.usda.gov/newsroom/2012/07/biotech_petitions.shtml
28. "USDA Petitions for Non-regulated Status Pending" http://www.aphis.usda.gov/biotechnology/not_reg.html Viewed July 24, 2012.
29. Pollack, Andrew. "Dow Weed Killer, Nearing Approval, Runs Into Opposition." The New York Times. April 25, 2012.
30. OPP Docket EPA. Physician Letter, June 22, 2012. Docket EPA–HQ–OPP–2011–0835 (corn).
31. "Monsanto pulls Roundup advertising in New York", *Wichita Eagle*, Nov. 27, 1996.
32. U.S. Congress. House of Representatives. Com. on Gov. Oper. 1984. *Problems plague the EPA pesticide registration activities.* House Report 98-1147.

Additional background resources for Diagnosis 5:

1. Strother, LeAnn. Tackling Tough Weeds. Glyphosate Resistance May Limit Control Options. October 2011. Soybeanreview.com.
2. Powles, Stephen. 2008. Evolved Glyphosate Resistant Weeds Around the World: Lessons to Be Learnt. Pest Management Science 64:360-365
3. Monaco, Veller, Ashton. 2002 Weed Science: Principles & Practices. John Wiley & Sons, Pg 326-334
4. Huber, Don. 2010. "What's New in Ag Chemical and Crop Nutrient Interactions." Fluid Journal: Official Journal of the Fluid Fertilizer Foundation. Volume 18 No 3 Issue 69.

http://www.scribd.com/Hooky1979/d/31835733-Glyphosate-101

5. András Székács and Béla Darvas. 2012. Forty Years with Glyphosate, Herbicides - Properties, Synthesis and Control of Weeds, Dr. Mohammed Nagib Hasaneen (Ed.), ISBN: 978-953-307-803-8. http://www.intechopen.com/books/herbicides-properties-synthesis-and-control-of-weeds/forty-years-with-glyphosate
6. Squibb, Katherine. "Applied Toxicology" Powerpoint. UM Program in Toxicology. Viewed 7/5/2012. https://docs.google.com/viewer?a=v&q=cache:d6GjqUYNx9gJ:aquaticpath.umd.edu/appliedtox/pesticides.pdf+examples+of+class+1+pesticides
7. US EPA. Glyphosate Ecological Risk Document. (EPA-HQ-OPP-2009-0361-0007) 7/22/2009.
8. U.S. EPA. Glyphosate Human Health Scoping Document (EPA-HQ-OPP-2009-0361-0007) 12/30/2009.

Diagnoses 6: Displaced Abomasa & Clostridials

1. http://www.motherearthnews.com/sustainable-farming/grass-fed-cattle-zm0z10zrog.aspx Viewed July 29, 2012.
2. *Rasby, R. Anderson, B. Randle, R. Beef* May 2010. University of Nebraska-Lincoln Extension. Bloat Prevention and Treatment in Cattle. http://www.ianrpubs.unl.edu/pages/publicationD.jsp?publicationId=1290
3. *U.*S. Patent #7771736. Viewed July 25, 2012.
4. *Kruger et al. 2011. The affect of Glyphosate on A. faecalis and C. botulinum.* Preliminary Data Chart.
5. Crees, Alex. *Rate of C. difficile infection increasing dramatically among children.* FoxNews.com. May 21, 2012. Viewed March 21, 2013. http://www.foxnews.com/health/2012/05/21/rate-c-difficile-infection-increasing-dramatically-among-children/
6. Kruger, M. et al. 2013. Glyphosate suppresses the antagonistic effect of Enterococcus spp. on Clostridium botulinum. Anaerobe. Feb 6, 2013. Doi: 10.1016/j.anaerobe.2013.01.005.
7. Rodloff AC, Kruger M. 2012. Chronic Clostridium botulinum infections in farmers, Anaerobe. doi: 10.1016/j.anaerobe.2011.011
8. E.C. Keessenet al. Clostridium difficile infection in humans and animals, differences and similarities. Veterinary Microbiology 153 (2011) 205–217
9. Marie Louise Hermann-Bank. July 31, 2012. National Veterinary Institute. English Translation: http://www.maskinbladet.dk/artikel/forskere-forsoger-finde-arsag-ny-spaedgri
10. "Soil study shows rise in resistance genes." *Feedstuffs*. January 4, 2010.
11. Huber, Don. Proceedings Fluid Fertilizer Forum, Scottsdale, AZ February 14-16, 2010. Vol. 27. Fluid Fertilizer Foundation, Manhattan, KS.
12. G.S. Johal, D.M. Huber / Europ. J. Agronomy 31 (2009) 144–152.

Diagnoses 7: Listeriosis

1. "Listeriosis." The Merch Veterinary Manual. http://www.realmilk.com/safety/safety-of-raw-milk Viewed Februray 5, 2013.
2. Linda Bren. 2004. "Got Milk? Make Sure It's Pasteurized." US Food and Drug Administration.
3. McAfee, Mark. http://www.realmilk.com/safety/safety-of-raw-milk/ Viewed March 4, 2013.
4. Rich, Robert. "Keeping it raw". *The Mountain View Voice* (Embarcadero Publishing Company). Sept. 5, 2003. Viewed July 12, 2012.
5. *Raw milk consumers say pasteurization is not needed*. CTV News, British Columbia. Viewed July 12, 2012.
6. "A Campaign for Real Milk/The Weston A. Price Foundation".
7. Stabel, J. R.; Lambertz, A. (2004), "Efficacy of Pasteurization Conditions for the Inactivation of *Mycobacterium avium* subsp. *paratuberculosis* in Milk", *Journal of Food Protection* 67 (12): 2719.
8. Capper, Jude. "Can buying local food really save the planet?" *Hoard's Dairyman*. January 25, 2012.
9. Gueber, Alan. "Feeding the world no longer possible with lagging exports." The Des Moines Register. Viewed March 13, 2013.
10. Thicke, Francis. A New Vision for Iowa Food and Agriculture. Mulberry Knoll Books, 2010, pg 158-159.
11. "How Much Sugar Do You Eat? You May be Surprised!" Viewed March 6, 2013. www.dhhs.nh.gov/DPHS/nhp/adults/documents/sugar.pdf
12. National Corn Growers Association. "Corn Statistics." Viewed July 2012.
13. National Sustainable Ag Coalition. *Path to the 2012 Farm Bill: Local Food Marker Bill: Where are we Now?* 7/31/2012. Viewed: August 6, 2012. http://sustainableagriculture.net/blog/path-to-the-2012-farm-bill-lffja-marker-bill-where-are-we-now/
14. EWP Report: Viewed August 8, 2012. http://static.ewg.org/reports/2012/farm_bill/babcock_free_crop_insurance.pdf
15. Adler, Liz. Mountain View Farm. Interview. Conducted August 13, 2012.

Diagnosis 8: Mycotoxicosis

1. Croddy, Eric. *Weapons of Mass Destruction.* Google Ebook, 2005. Pg 337.
2. Nijhuis, Michelle. Posted Oct. 24, 2011. *A Rise in Fungal Diseases is Taking a Toll on Wildlife.* http://e360.yale.edu/feature/a_rise_in_fungal_diseases_is_taking_growing_toll_on_wildlife/2457/
3. Koenning, Steve. Mycotoxins in Corn. Viewed March 18, 2013. http://www.ces.ncsu.edu/depts/pp/notes/Corn/corn001.htm
4. Hoskins, Tom. April 23, 2011. "Glyphosate effect beyond weed control gaining attention." Iowa Farmer today.

5. The Organic and Non-GMO report. Interview with Robert Kremer. 2010. http://www.non-gmoreport.com/articles/jan10/scientists_find_negative_impacts_of_GM_crops.php
6. Kremer, R.J., Means, N.E., 2009. Glyphosate and glyphosate-resistant crop interactions with rhizosphere microorganisms. Eur. J. Agron. 31, 153–161.
7. Kremer, R.J., Means, N.E., Kim, S.J., 2005. Glyphosate affects soybean root exudationand rhizosphere microorganisms. Int. J. Environ. Anal. Chem. 85, 1165–1174.
8. Kremer et al. *Glyphosate interactions with physiology, nutrition, and diseases of plants: Threat to agricultural sustainability?* Europ. J. Agronomy 31, 2009. 111–113.
9. http://www.greenpasture.org/utility/showArticle/?objectID=7213
10. http://farmindustrynews.com/fungicides/do-soybean-fungicides-pay
11. http://pdc.unl.edu/agriculturecrops/corn/gosswilt Viewed August 18, 2011.
12. Raymond, Richard. "How far back to ag basics do we go?" *Feedstuffs*. March 4, 2013.
13. Rastetter, Bruce. "Economic and Food Security is the AgriSol Goal in Tanzania." Jan 12, 2012. http://www.agrisoltanzania.com/news_files/fe1b5d870c752bb53ae54f25dbc17db-1.html
14. http://www.agrisoltanzania.com/kilele.html Viewed July 28, 2012.
15. www.oaklandinstitute.org/sites/oaklandinstitute.org/files/OI_brief_myths_and_facts_agrisol_energy_1.pdf, Viewed July 28, 2012.
16. Gueber, Alan. "What's going on with ISU, Rastetter?" *Agrinews*. Jan 12, 2012.
17. Krogstad, Jens. Group accuses Rastetter of ethics violations. Des Moines Register. June 20, 2012.
18. Iowa City Associated Press. "Hog, Ethanol Baron Now a Republican KingMaker. http://wcfcourier.com/news/local/govt-and-politics/hog-ethanol-baron-bruce-rastetter-now-a-republican-kingmaker/article_f464020c-84d8-5b04-8fb4-176173652e20.html Viewed Jan 2013.
19. American Rural Fund Website. Strategic Partner Biography. http://www.raflp.com/site/epage/90389_864.htm Viewed July 2012.
20. Gueber, Alan. "Good Old Farm boy Network is Alive and Well." Agrinews Commentary. January 19, 2012.
21. Mulvaney, Patrick. *Africa: GM Dumping-Ground*. http://www.organicconsumers.org/biod/africa050404.cfm, 2004. Viewed: March 2012.
22. *Pioneer acquisition of Pannar approved*. Feedstuffs Magazine. June 11, 2012. Pg 20.
23. Cooke, Jennifer and Downie, Richard. 2010. *African Perspectives on Genetically Modified Crops*. Center for Strategic and International Studies.
24. http://www.acbio.org.za/index.php/publications/gmos-in-south-africa/388 Viewed August 1, 2012.
25. http://triplecrisis.com/who-pays-for-agricultural-dumping-farmers-in-developing-countries-2/, Viewed January 2012.

26. Lynn, Jonathan (16 August 2009). "Doha deal could boost world GDP $300–700 billion: study". Reuters. Viewed July 2012.
27. Brown, Drusilla K.; Alan V. Deardorff and Robert M. Stern. December 2002. (PDF). *Computational Analysis of Multilateral Trade Liberalization in the Uruguay Round and Doha Development Round*. School of Public Policy. The University of Michigan.
28. Hertel, Thomas W.; Roman Keeney. 2005. "What is at Stake: The Relative Importance of Import Barriers, Export Subsidies and Domestic Support" (PDF). World Bank. Viewed July 2012.
29. Kym Anderson and Bjørn Lomborg, *Free Trade, Free Labor, Free Growth*, Project Syndicate. May, 2008.
30. Kenya faces grim prospects if Doha round succeeds at Corporate News in Business Daily, IPS report. Nairobi, Kenya. November 2009.
31. Rio+20 Summit. Food and Agriculture Organization. June 20-22, 2012.
32. Food and Agriculture Organization. *Save and Grow.* http://www.fao.org/ag/save-and-grow/index_en.html, Viewed July 29, 2012.
33. http://www.fao.org/ag/save-and-grow/en/6/index.html, Viewed July 30, 2012.
34. McPhee, Katharine. July 2002. *Neoliberalism on the molecular scale. Economic and genetic reductionism in biotechnology battles*. Geoforum 34 (2003) 203-219.

Diagnosis 9: Mulberry Heart Syndrome

1. U-Mass Extension. April 26, 2010. "Phosphorus Management in Pork Production." http://www.extension.org/pages/27471/phosphorus-management-in-pork-production
2. Dunham, Art. Interview. August 8, 2012.
3. The Economist. *Blooming Horrible*. June 23, 2012.
4. Forsberg, C.W., Phillips, J.P., Golovan, S.P., Fan, M.Z., Meidinger, R.G., Ajakaiye A., Hilborn D., and Hacker R.R. February 2003. "The Enviropig physiology, performance, and contribution to nutrient management, advances in a regulated environment: The leading edge of change in the pork industry". *Journal of Animal Science* 81 (14 Suppl 2): E68–E77.
5. Theriault A, Chao JT, Wang Q, Gapor A, Adeli K. July 1999. "Tocotrienol: a review of its therapeutic potential". *Clinical Biochemistry* 32 (5): 309–19. doi:10.1016/S0009-9120(99)00027-2. PMID 10480444.
6. Müller L, Theile K, Böhm V. May 2010. "In vitro antioxidant activity of tocopherols and tocotrienols and comparison of vitamin E concentration and lipophilic antioxidant capacity in human plasma". *Mol Nutr Food Res* 54 (5): 731–42. doi:10.1002/mnfr.200900399. PMID 20333724.
7. Yoshida Y, Niki E, Noguchi N. March 2003. "Comparative study on the action of tocopherols and tocotrienols as antioxidant: chemical and physical effects". *Chemistry and Physics of Lipids* 123 (1): 63–75. doi:10.1016/S0009-3084(02)00164-0. PMID 12637165.
8. Pearce BC, Parker RA, Deason ME, Qureshi AA, Wright JJ. October 1992. "Hypocholesterolemic activity of synthetic and natural tocotrienols". *Journal of Medicinal Chemistry* 35 (20): 3595–606. doi:10.1021/jm00098a002.

PMID 1433170.
9. Stocker A. December 2004. "Molecular mechanisms of vitamin E transport". *Ann. N. Y. Acad. Sci.* 1031: 44–59. doi:10.1196/annals.1331.005. PMID 15753133.
10. Ikeda S, Tohyama T, Yoshimura H, Hamamura K, Abe K, Yamashita K. February 2003. "Dietary alpha-tocopherol decreases alpha-tocotrienol but not gamma-tocotrienol concentration in rats". *J. Nutr.* 133 (2): 428–34. PMID 12566479.
11. Shibata A, Nakagawa K, Sookwong P, Tsuduki T, Asai A, Miyazawa T. June 2010. "alpha-Tocopherol attenuates the cytotoxic effect of delta-tocotrienol in human colorectal adenocarcinoma cells". *Biochem. Biophys. Res. Commun.* 397 (2): 214–9. doi:10.1016/j.bbrc.2010.05.087. PMID 20493172.
12. Sontag TJ, Parker RS. May 2007. "Influence of major structural features of tocopherols and tocotrienols on their omega-oxidation by tocopherol-omega-hydroxylase". *J. Lipid Res.* 48 (5): 1090–8. doi:10.1194/jlr.M600514-JLR200. PMID 17284776.
13. Abner, EL; Schmitt, FA, Mendiondo, MS, Marcum, JL, Kryscio, RJ (July 2011). "Vitamin E and all-cause mortality: a meta-analysis". *Current aging science* 4 (2): 158–70. PMID 21235492.
14. Miller Er, 3.; Pastor-Barriuso, R.; Dalal, D.; Riemersma, R. A.; Appel, L. J.; Guallar, E. (2005). "Meta-analysis: High-dosage vitamin E supplementation may increase all-cause mortality". *Annals of internal medicine* 142 (1): 37–46. PMID 15537682.
15. Bjelakovic, G; Nikolova, D, Gluud, LL, Simonetti, RG, Gluud, C. April 2008. Bjelakovic, Goran. ed. "Antioxidant supplements for prevention of mortality in healthy participants and patients with various diseases". *Cochrane database of systematic reviews (Online)* (2): CD007176. doi:10.1002/14651858.CD007176. PMID 18425980.
16. Bin, Q; Hu, X, Cao, Y, Gao, F. April 2011. "The role of vitamin E (tocopherol) supplementation in the prevention of stroke. A meta-analysis of 13 randomized controlled trials". *Thrombosis and haemostasis* 105 (4): 579–85. doi:10.1160/TH10-11-0729. PMID 21264448.
18. Haederle, Michael. "Vitamin E Supplements Raise Risk of Prostate Cancer". *Health Discovery*. AARP. Viewed November 2011.
19. Olson, JH; Erie, JC, Bakri, SJ (May 2011). "Nutritional supplementation and age-related macular degeneration". *Seminars in ophthalmology* 26 (3): 131–6. doi:10.3109/08820538.2011.577131. PMID 21609225.
20. "Taking vitamin E linked to osteoporosis: research". Viewed March 6, 2013.
21. Miller, E and Ullrey, D. Baby Pig Anemia. Pork Information Gateway FactSheet. April 1, 2007.
22. Vansickle, Joe. Setting Pigs Up to Succeed. Investing in the first day of a pig's life can make a pivotal difference in its lifetime performance. National Hog Farmer. Jun. 15, 2009. Viewed March 12, 2013. *http://nationalhogfarmer.com/genetics-reproduction/sow-gilt/0615-investing-first-day*
23. "Mulberry Heart Disease." *The Pig Site*. Viewed March 2013, 2013. http://www.thepigsite.com/diseaseinfo/69/mulberry-heart-disease-vitamin-e-deficiency
24. Vitamin E Deficiency and Iron Toxicity. *The Pig Site.*

http://www.thepigsite.com/pighealth/article/303/vitamin-e-deficiency-and-iron-toxicity. Viewed March 12, 2013.

25. Cromwell, Gary. Selenium: A Unique Trace Element. Professor, Swine Nutrition. Hand-out. Viewed March 15, 2013.
26. Vansickle, Joe. "Fixing Vitamin D Deficiency." Oral Supplementation helps ensure normal skeletal development. *National Hog Farmer*. Feb. 15, 2012. Viewed March 13, 2013.

Diagnosis 10: Lame Feet

1. Lameness in Dairy Cattle. University of Massachusetts Extension. January, 27, 2011. http://www.extension.org/pages/10855/lameness-in-dairy-cattle
2. Hoffman, A. Footbaths for the Treatment and Control of Hairy Heel Warts. Veterinary Medicine Extension, Washington State University. August 2012.
3. Sullivan, Hillary. "Hairy Foot Warts." New Mexico State Extension. 3/10/2013.
4. "Digitial Dermatitis." The Merck Veterinary Manual. http://www.merckvetmanual.com/mvm/index.jsp?cfile=htm/bc/90530.htm Viewed 3/12/2013.
5. http://horsetalk.co.nz/2012/09/27/protein-discoveries-systemic-nature-laminitis/#.UUcgc44pvloViewed: March 14, 2013.
6. http://www.aces.edu/pubs/docs/H/HE-0426/ Viewed: March 14, 2013.
7. Raymond, Richard. "Our food safety system isn't failing." *Feedstuffs*. July 9, 2012.
8. Federal Defense Budget, 2011.
9. The Union of Concerned Scientists. February 2012. "Heads they Win, Tails We Lose: How Corporations Corrupt Science At the Public's Expense."
10. Dagorret: Monsanto and its Seed Patents. Viewed July 2012. http://www.dagorret.net/monsanto-and-its-seed-patents-genetically-modified/
11. ISIS report: Monsanto versus Farmers. April 28, 2005. http://www.i-sis.org.uk/MonsantovsFarmers.php
12. Adam, David. GM crops: What are genetically modified crops, and should we be concerned about them? Viewed March 2013. www.guardian.co.uk/environment/2007/jul/26/gmcrops
13. USDA: ISB Data search, http://www.isb.vt.edu/data.aspx
14. TRIPs, Section 5, Article 27.
15. US patent #6,239,072 at http://www.uspto.gov/patft/index.html
16. Wallace, Richard. Hairy Heel Warts: Fads and Fashions. 7/27/1999.

Diagnosis 11: Herd Health

1. Sinclair, Upton. *I, Candidate for Governor and How I Got Licked*. University of California Press, 1994.

For more information visit www.americastwoheadedpig.com

Made in the USA
Lexington, KY
24 January 2015